AQA

GCSE

MATHEMATICS

A to A*

PRACTICE BOOK

Series editor: **Glyn Payne**

Author: **Harry Smith**

Longman

Part of Pearson

Longman is an imprint of Pearson Education Limited, a company incorporated in England and Wales, having its registered office at Edinburgh Gate, Harlow, Essex, CM20 2JE. Registered company number: 872828

www.pearsonschoolsandfecolleges.co.uk

Longman is a registered trademark of Pearson Education Limited

Text © Pearson Education Limited 2010

First published 2010
14 13 12 11 10
10 9 8 7 6 5 4 3 2 1

British Library Cataloguing in Publication Data
A catalogue record for this book is available from the British Library.
ISBN 978 1 408 24089 2

Edited by Sue Gardner
Designed by Pearson Education Limited
Typeset by Tech-Set Ltd, Gateshead
Original illustrations © Pearson Education Ltd 2010
Illustrated by Tech-Set Ltd
Cover design by Wooden Ark
Cover photo © Shutterstock/Mypokcik
Printed in the UK by Scotprint

Acknowledgements
Every effort has been made to contact copyright holders of material reproduced in this book. Any omissions will be rectified in subsequent printings if notice is given to the publishers.

Quick contents guide

Chapter Mapping	*vi*
Specification changes	*viii*
How to use this book	*ix*
Challenge yourself	*x*
Quality of written communication	*xi*
About the Digital Edition	*xii*

UNIT 1 Statistics and Number

1 Handling data	1	3 Proportionality	11
2 Probability	8	4 Accuracy in calculations	15

UNIT 2 Number and algebra   Non-calculator

5 Fractions and algebra	19	8 Quadratics	29
6 Indices and surds	21	9 Further algebraic methods	35
7 Algebraic methods	25		

UNIT 3 Geometry and algebra

10 Shape	40	15 Non-linear graphs	60
11 Applications of density and speed	45	16 Further trigonometry	69
12 Congruency and similarity	47	17 Transformations of graphs	76
13 Pythagoras' theorem	52	18 Vectors	80
14 Circle theorems	56		

Modular practice papers	*83*
Linear practice papers	*91*

Contents

Chapter Mapping   vi
Specification changes   viii
How to use this book   ix
Challenge yourself   x
Quality of written communication   xi
About the Digital Edition   xii

UNIT 1 Statistics and Number

1	Handling data	1
1.1	Stratified sampling	1
1.2	Drawing histograms	2
1.3	Interpreting histograms	4
1.4	Which average?	6

2	Probability	8
2.1	Tree diagrams	8
2.2	Conditional probability	9

3	Proportionality	11
3.1	Direct proportion	11
3.2	Other types of proportionality	12
3.3	Inverse proportion	13

4	Accuracy in calculations	15
4.1	Accuracy in calculations	15
4.2	Problem solving	16
4.3	Percentage error	17

UNIT 2 Number and Algebra

5	Fractions and algebra	19
5.1	Converting recurring decimals to fractions	19
5.2	Solving problems with simultaneous equations	20

6	Indices and surds	21
6.1	Simplifying powers and roots	21
6.2	Further use of indices	22
6.3	Surds	22
6.4	Rationalising the denominator	23
6.5	More complicated surds	24

7	Algebraic methods	25
7.1	Proof	25
7.2	Perpendicular lines	26
7.3	Velocity–time graphs	28

8	Quadratics	29
8.1	Factorising the difference of two squares	29
8.2	Factorising quadratics of the form $ax^2 + bx + c$	30
8.3	Solving equations by factorising	31
8.4	Using the quadratic formula	32
8.5	The discriminant	33
8.6	Completing the square	33

9	Further algebraic methods	35
9.1	Simultaneous equations… one linear, one quadratic	35
9.2	Simplifying algebraic fractions	36
9.3	Equations involving algebraic fractions	37
9.4	Changing the subject of a formula	38

UNIT 3 Geometry and Algebra

10	Shape	40
10.1	Pyramids	40
10.2	Sectors and arcs	42
10.3	Cones	43
10.4	Spheres	44

11	Applications of density and speed	45
11.1	Density and speed	45

12	**Congruency and similarity**	**47**
12.1	Areas and volumes of similar objects	47
12.2	Congruent triangle	49
12.3	Negative scale factors	51

13	**Pythagoras' theorem**	**52**
13.1	Pythagoras' theorem with pyramids and cones	52
13.2	Distances in 3-D	53
13.3	The angle between a line and a plane	54

14	**Circle theorems**	**56**
14.1	Proofs of circle theorems	56
14.2	The alternate segment theorem	58

15	**Non-linear graphs**	**60**
15.1	Solving quadratic equations graphically	61
15.2	Solving problems using quadratic equations	64
15.3	Graphs of reciprocal functions and combined functions	65
15.4	Graphs of exponential functions	66

16	**Further trigonometry**	**69**
16.1	Solving trigonometric equations	69
16.2	Calculating areas using trigonometry	71
16.3	The sine rule	72
16.4	The cosine rule	73
16.5	Solving 2-D and 3-D problems	75

17	**Transformations of graphs**	**76**
17.1	Transformations of graphs	76
17.2	Transformations of trigonometric graphs	78

18	**Vectors**	**80**
18.1	Vectors	80
18.2	Vector geometry	81

Modular practice papers		*83*
Linear practice papers		*91*

rades A to A* ... Grades A to A* ... Grades A to A* ... Grades A to A* ...

v

This book has been written to match Longman's **AQA GCSE Maths Higher Sets Student Book**. It contains extra practice for every A and A* topic in the student book.

AQA GCSE Maths Higher Sets Student Book Chapter	A-A* Practice Book sections	A-A* Practice Book page numbers
1 Data collection	1 Handling data	1–2
2 Fractions, decimals and percentages	—	—
3 Interpreting and representing data	1 Handling data	2–6
4 Range and averages	1 Handling data	6–7
5 Probability	2 Probability	8–10
6 Cumulative frequency	—	—
7 Ratio and proportion	3 Proportionality	11–14
8 Complex calculations and accuracy	4 Accuracy in calculations	15–18
9 Estimation and currency conversion	—	—
10 Factors, powers and roots	—	—
11 Fractions	—	—
12 Basic rules of algebra	—	—
13 Decimals	5 Fractions and algebra	19
14 Equations and inequalities	5 Fractions and algebra	20
15 Formulae	—	—
16 Indices and standard form	6 Indices and surds	21–24
17 Sequences and proof	7 Algebraic methods	25–26
18 Percentages	—	—
19 Linear graphs	7 Algebraic methods	26–28
20 Quadratic equations	8 Quadratics	29–34

AQA GCSE Maths Higher Sets Student Book Chapter *(cont...)*	A-A* Practice Book sections *(cont...)*	A-A* Practice Book page numbers *(cont...)*
21 Further algebra	9 Further algebraic methods	35–39
22 Number skills	—	—
23 Angles	—	—
24 Triangles, polygons and constructions	—	—
25 More equations and formulae	—	—
26 Compound shapes and 3-D objects	10 Shape	40–41
27 Circles, cylinders, cones and spheres	10 Shape	42–44
28 Measures and dimensions	11 Applications of density and speed	45–46
29 Constructions and loci	—	—
30 Reflection, translation and rotation	—	—
31 Enlargement	—	—
32 Congruency and similarity	12 Congruency and similarity	47–51
33 Pythagoras' theorem and trigonometry	13 Pythagoras' theorem	52–55
34 Circle theorems	14 Circle theorems	56–59
35 Non-linear graphs	15 Non-linear graphs	60–68
36 Further trigonometry	16 Further trigonometry	69–75
37 Transformations of graphs	17 Transformations of graphs	76–79
38 Vectors	18 Vectors	80–82

From 2010 the AQA GCSE Maths specifications have changed. For both Modular and Linear, the main features of this change are twofold.

Firstly the Assessment Objectives (AOs) have been revised so there is more focus on problem-solving. The new AO2 and AO3 questions will form about half of the questions in the exam. We provide lots of practice in this book, with AO2 and AO3 questions clearly labelled.

Secondly about a quarter of the questions in the exam will test functional maths. This means that they use maths in a real-life situation. Again we provide lots of clearly labelled practice for functional questions.

What does an AO2 question look like?

"**AO2** select and apply mathematical methods in a range of contexts."

An AO2 question will ask you to use a mathematical technique in an unfamiliar way.

A

5 The values given in this newspaper article are correct to 2 significant figures. Calculate the maximum and minimum values for the mean amount raised for charity by each runner.

AO2

8500 RUN IN HALF MARATHON
£1.2 MILLION RAISED FOR CHARITY

> Calculate maximum and minimum values for the number of runners and amount of money raised.

> Use these values to calculate the maximum mean and the minimum mean. Simple!

What does an AO3 question look like?

"**AO3** interpret and analyse problems and generate strategies to solve them."

AO3 questions give you less help. You might have to use a range of mathematical techniques, or solve a multi-step problem without any guidance.

A*

2 OPQ is a triangle.

$OM : MP = ON : NQ = 1 : 2$. $\overrightarrow{OM} = \mathbf{a}$ and $\overrightarrow{ON} = \mathbf{b}$.

a   Write expressions for $\overrightarrow{MN}$ and $\overrightarrow{PQ}$ in terms of $\mathbf{a}$ and $\mathbf{b}$.

AO3

b   What does your answer to part a tell you about the lines MN and PQ?

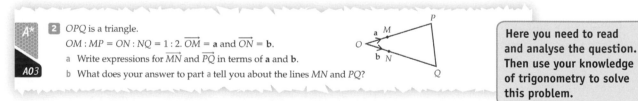

> Here you need to read and analyse the question. Then use your knowledge of trigonometry to solve this problem.

What does a functional question look like?

When you are answering functional questions you should plan your work. Always make sure that you explain how your answer relates to the question.

A

5 A paint manufacturer makes three different shades of green.

APPLE WHITE RATIO WHITE : GREEN **4:1**

HINT OF MINT RATIO WHITE : GREEN **7:3**

SUMMER FIELDS RATIO WHITE : GREEN **1:1**

Hamish has run out of *Hint of Mint*, and wants to mix *Apple White* with *Summer Fields* to create more. How much of each type of paint should Hamish use to mix 15 litres of *Hint of Mint*?

> Read the question carefully.

> Think what maths you need and plan the order in which you'll work.

> Follow your plan. Check your calculation. Job done!

How to use this book

This book has all the features you need to achieve the best possible grade in your AQA GCSE Higher exam, **both Modular and Linear**. Throughout the book you'll find full coverage of Grades A-A*, the new Assessment Objectives and Functional Maths.

At the end of the book you will find a **complete set of Practice Papers for Modular and a complete set for Linear.**

Key points at the start of every chapter – a quick reminder of the main skills you'll need for that chapter.

Links to the Higher Sets student book in case you need extra help.

Challenge yourself sections at the end of every exercise – a chance to show off with a trickier problem.

Examiners' hints when you really need them.

All questions graded, with A02, A03 and Functional questions clearly indicated.

Every question graded, with A02 & A03 clearly highlighted – plenty of opportunities to practise your problem solving skills.

5 Practice Papers at back of book: complete set of Modular (Units 1-3) and complete set of Linear (Papers 1 & 2)

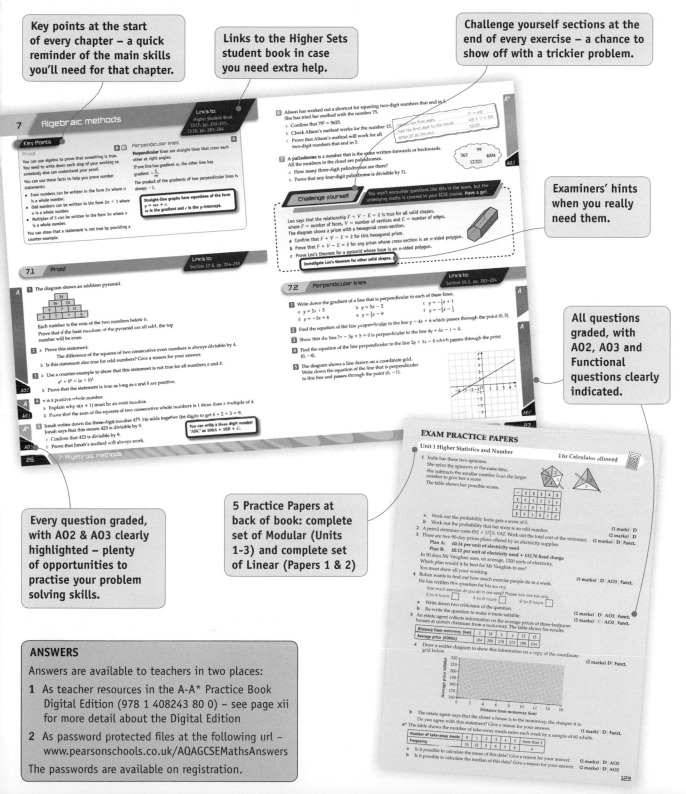

ANSWERS

Answers are available to teachers in two places:

1 As teacher resources in the A-A* Practice Book Digital Edition (978 1 408243 80 0) – see page xii for more detail about the Digital Edition

2 As password protected files at the following url – www.pearsonschools.co.uk/AQAGCSEMathsAnswers

The passwords are available on registration.

At the end of every exercise you'll find a **Challenge yourself** box. These problems use the same maths you've been practising in the exercise, but are tougher!

Some are a little bit harder than the questions you'll see on your GCSE exam, some are a lot harder. We have done this to provide genuine stretch-and-challenge, to uncover the excitement of mathematics, and to provide talking points for class discussions. So if you can solve them, you know you've got that topic cracked. Flex your mathematical muscles and have a go.

Into the unknown

Don't panic if the problem seems unfamiliar. You will only need to use the maths you have already practised in the exercise, or earlier in the book. If there is a hint then you could use that as a starting point. Try writing down the information given in the question or drawing a new diagram – remember mathematicians think on paper, not in their heads!

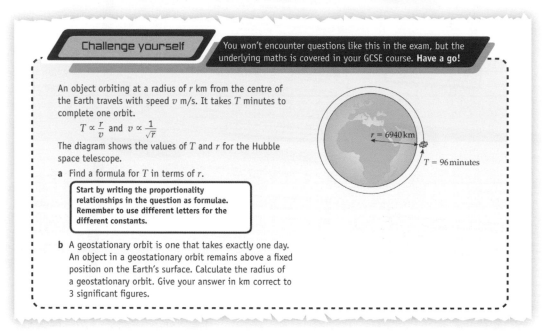

Challenge yourself

You won't encounter questions like this in the exam, but the underlying maths is covered in your GCSE course. **Have a go!**

An object orbiting at a radius of r km from the centre of the Earth travels with speed v m/s. It takes T minutes to complete one orbit.

$$T \propto \frac{r}{v} \quad \text{and} \quad v \propto \frac{1}{\sqrt{r}}$$

The diagram shows the values of T and r for the Hubble space telescope.

a Find a formula for T in terms of r.

> Start by writing the proportionality relationships in the question as formulae. Remember to use different letters for the different constants.

b A geostationary orbit is one that takes exactly one day. An object in a geostationary orbit remains above a fixed position on the Earth's surface. Calculate the radius of a geostationary orbit. Give your answer in km correct to 3 significant figures.

$r = 6940\,\text{km}$

$T = 96\,\text{minutes}$

Investigate!

Some of the Challenge yourself boxes are investigations. There might be more than one way to approach the problem. If the problem asks you to find a general solution to a problem try substituting some values first – this can give you a really good idea of what is going on.

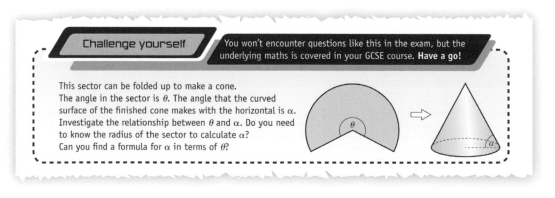

Challenge yourself

You won't encounter questions like this in the exam, but the underlying maths is covered in your GCSE course. **Have a go!**

This sector can be folded up to make a cone.
The angle in the sector is θ. The angle that the curved surface of the finished cone makes with the horizontal is α.
Investigate the relationship between θ and α. Do you need to know the radius of the sector to calculate α?
Can you find a formula for α in terms of θ?

Quality of written communication

Some of the questions on your exam paper are marked with a star *. This means you can win a bonus mark for showing the **quality of your written communication**. The basic rule of thumb is to make sure the examiner can understand *everything* you write on your exam paper, not just your answer. You can achieve this by:

- writing neatly
- laying out your working clearly
- using the correct mathematical notation and vocabulary
- setting out your answer in logical steps.

SAMPLE ANSWER

*7 A company designs a kitchen bin in the shape of a cylinder and a hemisphere.

They claim that the bin has a volume of 40 litres.

Do you agree with this claim? You must show your workings.

CYLINDER   SPHERE

31808.6   14137.2

$31808.6 + \frac{14137.2}{2} = 38877.2 \div 1000 = 38.9$

YES

(4 marks)

> Although this student has worked out the volume of the bin correctly she will not get the bonus mark for quality of written communication.

> You must write the correct units at each stage of your working.

> If you need to use an answer in a new calculation start a new line of working.

IMPROVED ANSWER

*7 A company designs a kitchen bin in the shape of a cylinder and a hemisphere.

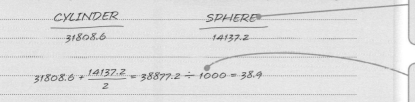

They claim that the bin has a volume of 40 litres.

Do you agree with this claim? You must show your workings.

$\pi \times 15^2 \times 45 = 31808.6 \text{ cm}^3$

$\frac{1}{2} \times \frac{4}{3} \times \pi \times 15^3 = 7068.6 \text{ cm}^3$

Total volume = 38877.2 cm³

38877.2 cm³ = 38.9 litres (1 d.p.)

The claim is reasonable

(4 marks)

> This student gets full marks. Her handwriting is neat and she has laid out her working logically.

> Write down the degree of accuracy you are using in your answer.

> It is always a good idea to answer the question with a sentence.

Grades A to A* . . . Grades A to A* . . . Grades A to A* . . . Grades A to A* . . .

XI

We have produced a **Digital Edition** of the A-A* Practice Book (ISBN 978 1 408243 80 0) for display on an electronic whiteboard or via a VLE. The digital edition is available for purchase separately. It makes use of our unique **ActiveTeach** platform and will integrate with any other ActiveTeach products that you have purchased from the **AQA GCSE Mathematics 2010 series**.

Complete flexibility: use the digital edition to display the Practice Book on a whiteboard or through a VLE.

Print out any page required from the bank of PDFs saved on the disc.

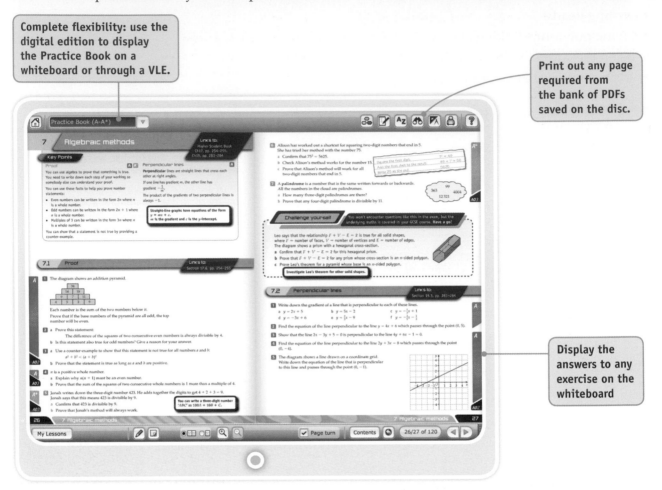

Display the answers to any exercise on the whiteboard

Higher Tier Resources in the AQA GCSE Mathematics 2010 Series

STUDENT BOOK	PRACTICE BOOK	EXTENSION PRACTICE BOOK	TEACHER GUIDE with EDITABLE CD-ROM
D-A* 9781408232781	D-A* 9781408232774	A-A* 9781408240892	D-A* 9781408232798

ACTIVETEACH CD-ROM	PRACTICE BOOK - Digital Edition	EXTENSION PRACTICE BOOK - Digital Edition	ASSESSMENT PACK with EDITABLE CD-ROM - Covering all sets
D-A* 9781408232767	D-A* 9781408243831	A-A* 9781408243800	G-A* 9781408232842

Links to:
Higher Student Book
Ch1 pp. 15–16, Ch3
pp. 45–48, Ch4 pp. 60–62

Key Points

Stratified sampling   **A** **A***

If a population is divided into groups, you can take a stratified sample. The **sampling fraction** is the total sample size written as a fraction of the population size. The number of people chosen from each group is the group size multiplied by the sampling fraction.

Histograms   **A** **A***

A histogram represents continuous data.

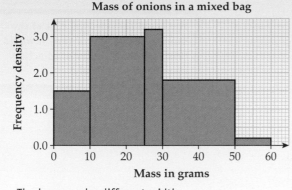

Mass of onions in a mixed bag

- The bars can be different widths.
- There are no gaps between the bars.
- The area of each bar is proportional to the frequency.
- The vertical axis represents frequency density.

$$\text{Frequency density} = \frac{\text{frequency}}{\text{class width}}$$

Appropriate averages   **A**

- The mode is the only average you can use for qualitative data. The mode tells you which data value is most likely to occur.

- The median tells you the middle value. Half the data values are greater than the median and half are less than the median.

- The mean is the only average that takes every value into account. The mean can be affected by very high or very low values. These values are called **extreme values**.

1.1   Stratified sampling

Links to:
Section 1.8, pp. 15–16

1 A primary school has 800 pupils in Years 1 to 6.

The headmistress needs to select a stratified sample of 100 pupils.

Year 1	Year 2	Year 3	Year 4	Year 5	Year 6
122	156	88	145	110	179

a   Write down the sampling fraction for this sample.

b   Calculate the number of pupils the headmistress should select from each year group.

2 A factory has five machines producing electrical components.
The table shows the number of components produced by each machine in one day.

Machine A	Machine B	Machine C	Machine D	Machine E
1260	2015	765	930	1850

The factory manager wants to take a 2% stratified sample of these components.
Work out how many components he should take from each machine.

3 This table shows the numbers of employees at a large department store.
A stratified sample of 40 employees is required.

	Full-time	Part time
Male	80	36
Female	105	19

a   Calculate the number of male employees that should be included in the stratified sample.

b   How many female full-time employees will be selected?

A

AO2

4 A company is surveying its workers about working conditions.
The company wants its survey to reflect the distribution of workers in its different offices.

Slough	Middlesbrough	Liverpool	Swansea
225	162	54	118

   a Calculate the number of workers from each office that should be selected for a 5% stratified sample.

   b Suggest one method the company could use to select a random sample.

A

5 This table shows the populations of three different towns.
A marketing survey is being carried out across the three towns. The survey will use a stratified sample of 250.

Woodbridge	Ipswich	Felixstowe
11 000	139 000	29 000

   a Calculate the number of people from each town that should be included in the sample.

   b Explain why it would be easier to use a stratified sample for this survey than to select a random sample from the entire population.

> **You could consider the cost of the sample, the length of time it would take, or the difficulty of selecting respondents randomly.**

AO3

A*

6 A football team supporters' club has three tiers of membership. This table shows the number of members in each tier.

	Bronze	Silver	Gold
Male	3126	6790	1480
Female	2290	5208	264

The club wants to select a stratified sample of 100 supporters.

   a Calculate the total number of silver members in the sample.

   b Calculate the numbers of male silver members and female silver members in the sample separately.

   c Explain why your answer to part a is not the sum of your answers to part b.

   d Would you select an additional male or female respondent from the silver members? Justify your answer.

AO3

Challenge yourself   You won't encounter questions like this in the exam, but the underlying maths is covered in your GCSE course. **Have a go!**

A secondary school is using a 6% stratified sample to help them choose a new school motto.
They select eight Year 7 students for the sample.
Calculate the maximum and minimum possible numbers of Year 7 students in the school.

1.2   **Drawing histograms**

Links to:
Section 3.5, pp. 45–48

A

1 Angelo measured the heights of some tree saplings for a biology project. He recorded his results in a grouped frequency table.

   a Copy and complete the frequency table.

   b Draw a histogram to represent this data.

> **Draw the vertical scale from 0 to 1 in steps of 0.1.**

Height, h (cm)	Frequency	Class width	Frequency density
$0 \leqslant h < 20$	5	20	0.25
$20 \leqslant h < 30$	6		
$30 \leqslant h < 40$	8		
$40 \leqslant h < 80$	8		

2 Jenny asked a group of university students about their weekly rent, and recorded her results in a frequency table.

Weekly rent, r	Frequency	Class width	Frequency density
£0 $\leq r <$ £50	9		
£50 $\leq r <$ £80	18		
£80 $\leq r <$ £100	32		
£100 $\leq r <$ £120	37		
£120 $\leq r <$ £160	38		

a Copy and complete the frequency table.

b Draw a histogram to represent this data.

3 This grouped frequency table shows the times taken by the members of a class to run a 100 m race.

Draw a histogram to represent this data.

Time taken, t (s)	Frequency
10 $\leq t <$ 12	3
12 $\leq t <$ 13	5
13 $\leq t <$ 14	6
14 $\leq t <$ 15	8
15 $\leq t <$ 20	12

4 Lydia tested the reaction times of the boys and girls in her year group. She recorded her results in a grouped frequency table.

Reaction time, t (s)	Boys' frequency	Girls' frequency
0 $\leq t <$ 0.2	8	10
0.2 $\leq l <$ 0.25	10	14
0.25 $\leq t <$ 0.3	13	9
0.3 $\leq t <$ 0.4	11	7
0.4 $\leq t <$ 0.5	7	8

a Draw a histogram to represent the boys' results.

b Draw a histogram to represent the girls' results.

c Write a statement comparing the boys' reaction times with the girls' reaction times. Use evidence from your histograms to support your statement.

Ahmet surveyed the waiting times of patients at a doctors' surgery for one week. He recorded his results in a grouped frequency table. One of the data values is missing.
Ahmet drew a histogram of his results. The bar representing the class 2 $\leq t <$ 5 was 1.5 cm wide and 1.9 cm high.

a Calculate the height and width of the bar representing the class
  i 0 $\leq t <$ 2
  ii 5 $\leq t <$ 10

b The bar representing the class 15 $\leq t <$ 30 had a height of 0.06 cm. Calculate the missing frequency in the table.

Waiting time, t (minutes)	Frequency
0 $\leq t <$ 2	48
2 $\leq t <$ 5	114
5 $\leq t <$ 10	170
10 $\leq t <$ 15	91
15 $\leq t <$ 30	□

A

1 This is a histogram of the diameters of tree trunks in an orchard.

Calculate the number of trees

a with a diameter of between 10 cm and 20 cm

b with a diameter of between 20 cm and 30 cm

c in the entire orchard.

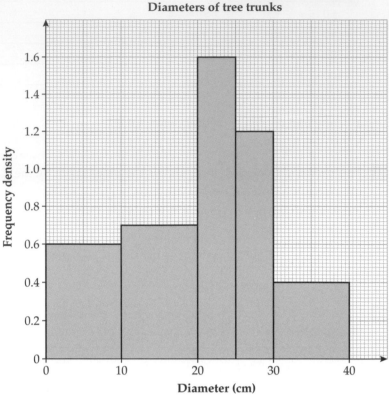

Diameters of tree trunks

A

2 This histogram shows the batting averages of the members of England's 2009 Ashes cricket squad.
The vertical axis has not been numbered.

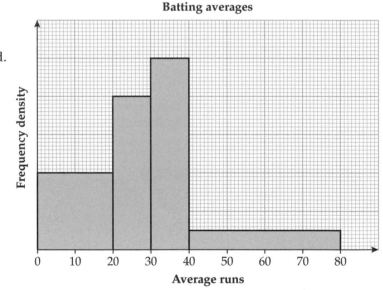

Batting averages

a Copy and complete this grouped frequency table of the data shown on the histogram.

Average runs, x	Frequency
$0 \leqslant x < 20$	4
$20 \leqslant x < 30$	
$30 \leqslant x < 40$	
$40 \leqslant x < 80$	

Calculate the height of the first bar to work out the scale on the vertical axis.

A02

b How many cricketers had a batting average of less than 40 runs?

3 This histogram shows the weights of the chickens for sale in a supermarket.

Work out an estimate for the number of chickens that weighed

a between 1.5 kg and 2 kg

b more than 2.75 kg.

c Explain why your answers to parts a and b are estimates.

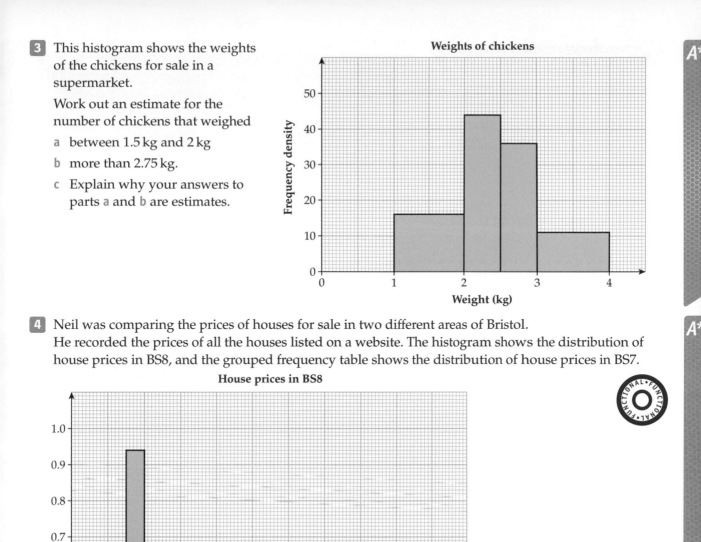

Weights of chickens

4 Neil was comparing the prices of houses for sale in two different areas of Bristol.
He recorded the prices of all the houses listed on a website. The histogram shows the distribution of house prices in BS8, and the grouped frequency table shows the distribution of house prices in BS7.

House prices in BS8

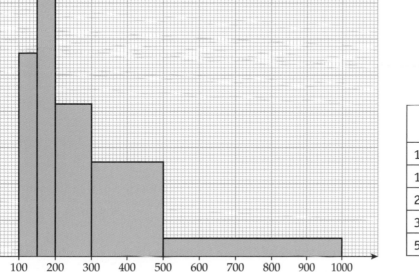

House prices in BS7

House price, x (£1000s)	Frequency
$100 \leqslant x < 150$	36
$150 \leqslant x < 200$	25
$200 \leqslant x < 300$	8
$300 \leqslant x < 500$	6
$500 \leqslant x < 1000$	5

a Estimate the number of houses for sale in BS8 for £400 000 or more.

b Draw a histogram to show the distribution of house prices in BS7.

c Write a sentence comparing the distribution of house prices in the two areas.

d In his conclusion Neil wrote,

'There were more houses for sale under £200 000 in BS8 than in BS7.
This suggests that houses in BS8 were cheaper than houses in BS7.'

Do you agree with Neil's conclusion? Give a reason for your answer.

Jonah recorded the weights of the apples in a box. This is the first line of his grouped frequency table.

Weight, w (g)	Frequency
$100 \leqslant w < 120$	6

On his histogram the bar for the class interval $100 \leqslant w < 120$ was 2 cm wide and 1.8 cm tall. The total area of all the bars in Jonah's histogram was 45 cm². How many apples were in the box?

1.4 Which average?

Links to:
Section 4.4, pp. 60–62

A

1 Andrew rolls a biased dice 50 times and records the results in a table.
Which average should he use to predict the next roll? Give a reason for your answer.

2 Gillian wrote down the prices of eight different pairs of trainers to the nearest pound.

£45   £49   £90   £52   £135   £55   £50   £90

In her conclusion Gillian wrote, 'An average pair of trainers costs £90'.

a Calculate the mean, median and mode of Gillian's data.

b Which average has Gillian used for her conclusion?

c Do you agree with Gillian's conclusion? Give a reason for your answer.

d Choose an appropriate average and write your own conclusion for Gillian's data.

3 Choose the most appropriate average for each investigation. Give reasons for your answers.

a Calculating the average salary of an employee at a company.

b Determining the most popular meal in the school canteen.

c Predicting the number of sweets in a tube.

d Comparing the densities of different types of wood.

A

4 For a biology experiment Heather measured the volume of water collected in six different test tubes.

23.1 ml 18.5 ml 17.8 ml 20.5 ml 1.8 ml 21.4 ml

a Write down the extreme value in Heather's data.

b Which average should Heather use if she doesn't want her data to be heavily influenced by the extreme value? Give a reason for your answer.

5 This is an extract from a sales report produced by a publishing company:

> Each copy of our *Super TV* listings magazine is read by an average of 3.5 people.

a Which two averages could the company have been using for this conclusion? Give reasons for your answers.

b Which average should the company use if they want to be able to calculate accurately the total readership of the magazine?

A02

6 This bar chart shows the ages of all the people who died in the United Kingdom last year.

The labels **A**, **B** and **C** represent the three averages.

a Which label represents which average?
Give reasons for your answers.

b Janet claims that most people live longer than the mean age.
Does this bar chart support her statement?
Give a reason for your answer.

c Which average would you use to represent the life expectancy of a typical person?
Give a reason for your answer.

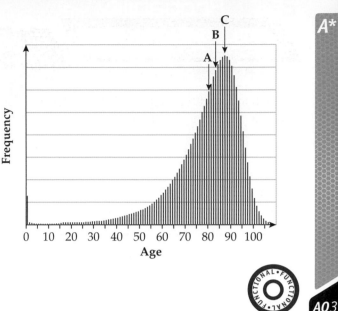

Challenge yourself

You won't encounter questions like this in the exam, but the underlying maths is covered in your GCSE course. **Have a go!**

Three classes are taking part in a school cross-country run.
The mean times and the numbers of students in each class are recorded in this table.

Class	Number of students	Mean time (min)
8A	24	41.0
8B	31	44.7
8C	28	36.5

Calculate the mean time taken by all the students who completed the run.
Give your answer correct to 1 decimal place.

Links to:
Higher Student Book
Ch5, pp. 81–88

Key Points

Tree diagrams   **A** **A***

A tree diagram is used to show the possible outcomes of two or more combined events.

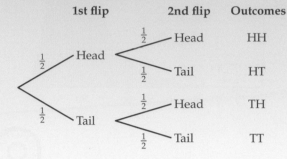

To find the probability of a combined outcome you multiply along the branches:

$$P(HH) = \tfrac{1}{2} \times \tfrac{1}{2}$$

To find the probability that either of two outcomes will occur you add their probabilities:

$$P(HT \text{ or } TH) = \tfrac{1}{4} + \tfrac{1}{4} = \tfrac{1}{2}$$

Conditional probability   **A***

If you pick one card from a pack of playing cards then

$$P(\text{Black}) = \tfrac{1}{2}$$

There are now 25 black cards in the pack and 26 red cards. If you pick a second card then

$$P(\text{Black}) = \tfrac{25}{51}$$

You can use tree diagrams to solve problems involving conditional probability.

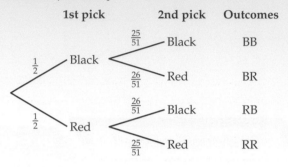

2.1   Tree diagrams

Links to:
Section 5.5, pp. 81–85

A

1 Jasmine has added modelling clay to one side of a coin. The probability that the coin lands on heads is $\tfrac{1}{4}$.

   a Copy and complete this tree diagram showing the possible outcomes of two flips of Jasmine's coin.

   b Calculate P(HH).

   c Work out the probability that Jasmine flips exactly one head.

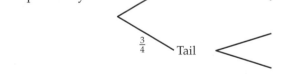

2 Karl creates a game where you spin a fair three-sided spinner then flip a fair coin.

   a Copy and complete this tree diagram showing all the possible outcomes.

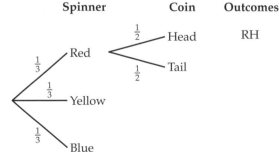

   b You win the game if the spinner lands on yellow or the coin lands on heads. Calculate the probability of winning the game.

3 The probability that it will rain in the morning is 0.4.
The probability that it will rain in the afternoon is 0.5.

| | Morning | Afternoon | Outcomes |

a Copy and complete this tree diagram to show all the possible outcomes.

b Work out the probability that it will rain in the morning and in the afternoon.

c If there is more than a 60% chance of rain, Adrian will take an umbrella. Should Adrian take an umbrella with him? Give a reason for your answer.

Morning *Afternoon* *Outcomes*

0.4 — Rain — Dry RD

— Dry

4

> EIGHT OUT OF TEN DOGS LOVE NEW CHEWALOT CHUNKS!

Three dogs are chosen at random. Work out the probability that

a none of them love Chewalot chunks

b exactly two of them love Chewalot chunks

c two or more of them love Chewalot chunks.

5 At a manufacturing plant, the probability of a component being faulty is 2%.
In one hour, three components are manufactured. Work out the probability of

a no component being faulty

b exactly one component being faulty

c at least one component being faulty.

Give your answers correct to 3 significant figures.

6 James, Emma and Imran have end-of-term tests in three subjects. Their probabilities of passing the different tests are shown in this table. Work out the probability that

a all three students pass their English test

b Emma fails all three tests

c Imran passes at least one test

d at least one student passes all three tests.

	Maths	English	Science
James	0.8	0.4	0.7
Emma	0.8	0.6	0.9
Imran	0.3	0.9	0.5

Challenge yourself

You won't encounter questions like this in the exam, but the underlying maths is covered in your GCSE course. **Have a go!**

A basketball player takes two shots from different places on the court. The probability that he scores with his first shot is 0.8. The probability that he scores with both shots is 0.6. Calculate the probability that the basketball player scores with at least one shot.

2.2 Conditional probability

Links to:
Section 5.6, pp. 86–88

1 A variety pack of yogurts contains 3 strawberry yogurts and 5 apricot yogurts.

Soujit eats two yogurts from the variety pack.

a Copy and complete this tree diagram to show the possible outcomes.

b Calculate P(SS).

c Work out the probability that Soujit eats two different-flavoured yogurts.

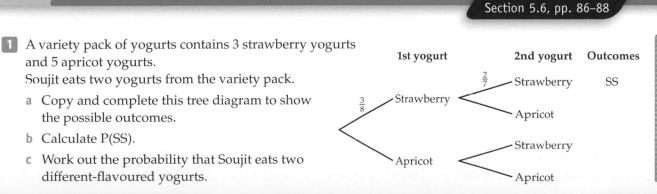

1st yogurt *2nd yogurt* *Outcomes*

$\frac{3}{8}$ — Strawberry — $\frac{2}{7}$ Strawberry SS

— Apricot

— Apricot — Strawberry

— Apricot

2 A pencil case contains 4 blue pens and 7 red pens. Chloe selects two pens at random.

   a Draw a tree diagram to show the possible outcomes.

   b What is the probability that Chloe selects two pens of the same colour?

3 Two counters are taken from this bag at random.

Work out the probability that

   a both counters are blue

   b at least one counter is yellow

   c the two counters are different colours.

4 On a game show there are nine boxes. Three boxes contain a prize. A contestant chooses three boxes at random. Work out the probability that the contestant will win

   a at least one prize

   b all three prizes

   c a prize with her final choice of box.

> Write down all the possible outcomes in which the contestant wins at least one prize, then work out their probabilities.

5 The probability that Will passes his driving test on the first attempt is 0.7.
If he fails, the probability that he passes it on his second attempt is 0.8.
Work out the probability that Will passes his driving test within two attempts.

6 Hayley's wallet contains these coins:

Hayley chooses three coins at random.
Work out the probability that her total is

   a 25p   b 80p

   c more than £1   d less than 50p.

7 Each day for school Alison chooses between wearing a skirt and wearing trousers. The probability that she will choose trousers is 0.6. If she chooses trousers, then the probability that she will wear trainers is 0.4. If she chooses a skirt, the probability that she will wear trainers is 0.2. Calculate the probability that

   a Alison does not wear trainers to school

   b in a five-day week, Alison will wear trainers exactly once.

8 A bag contains blue beads and red beads. Two beads are chosen at random. The probability of picking a blue bead is $\frac{1}{3}$. The probability of picking two blue beads is $\frac{1}{10}$. Work out

   a the number of blue beads initially in the bag

   b the total number of beads initially in the bag.

Challenge yourself

You won't encounter questions like this in the exam, but the underlying maths is covered in your GCSE course. **Have a go!**

In a lottery game, contestants select six numbers between 1 and 49. The numbers 1 to 49 are written on balls placed in a machine. Six of the balls are chosen at random. Calculate the probability of matching

a all six numbers

b exactly five numbers

c no numbers.

Give your answers to 3 significant figures.

Key Points

Direct proportion   **A**

Two quantities y and x are in direct proportion if the ratio between them stays the same as they are increased or decreased. You can write:

- y is directly proportional to x
- $y \propto x$
- $y = kx$ for some constant value k.

> **The value k is called the constant of proportionality.**

You can solve problems involving other types of proportionality by writing equations:

- $y \propto x^2$ $y = kx^2$
- $y \propto x^3$ $y = kx^3$
- $y \propto \sqrt{x}$ $y = k\sqrt{x}$

> $\propto$ **means 'is proportional to'**

Inverse proportion   **A**

Two quantities y and x are inversely proportional if their product is constant. You can write:

- y is inversely proportional to x
- $y \propto \dfrac{1}{x}$
- $y = \dfrac{k}{x}$ for some constant value k.

> **This equation can also be written as $yx = k$.**

You can use equations to solve problems involving other types of inverse proportionality:

- $y \propto \dfrac{1}{x^2}$ $y = \dfrac{k}{x^2}$
- $y \propto \dfrac{1}{x^3}$ $y = \dfrac{k}{x^3}$
- $y \propto \dfrac{1}{\sqrt{x}}$ $y = \dfrac{k}{\sqrt{x}}$

3.1   Direct proportion

Links to:
Section 7.7, pp. 118–120

1 Two variables are connected by the relationship $y \propto x$. When $x = 0.2$, $y = 1.8$.

   a Write a formula for y in terms of x.

   b Work out y when $x = 0.15$.

   c Work out x when $y = 27$.

   A

2 The cost of the petrol for a journey is directly proportional to the distance travelled. Graham travels 125 miles and spends £14 on petrol. Calculate

   a the cost of petrol for a journey of 160 miles

   b the length of a journey costing £10.50.

3 The cost of chocolate is proportional to the weight of the bar.

Weight (g)	81	189		
Cost	45p		£1.20	£2.22

Copy and complete this table showing the costs of different bars of chocolate.

4 Tania carries out an experiment comparing the temperature of a gas with its volume. She records the results in the table.

Temperature (K)	300	340	280
Volume (cm³)	187.5	212.5	175

   a Show that Tania's results are consistent with a direct proportionality relationship.

   b Does this mean that temperature is definitely proportional to volume? Give a reason for your answer.

   c Predict the volume of this gas at a temperature of 390 K.

   A

   AO2

A

5 The time taken for a kettle to boil, t, is directly proportional to the amount of water in the kettle, w. A kettle containing 600 ml of water takes 3 minutes and 20 seconds to boil.

  a Find a formula for t in terms of w.

  b How does the choice of units affect your formula?

  c How long will a kettle containing 1.2 litres of water take to boil?

  d How much water is in a kettle which takes 90 seconds to boil?

> Choose which units you are going to use for t and w.

A

AO2

AO3

6 If $a \propto b$ and $b \propto c$, show that $a \propto c$.

> ## Challenge yourself
>
> You won't encounter questions like this in the exam, but the underlying maths is covered in your GCSE course. **Have a go!**
>
> Paper comes in a range of standard sizes. If you cut a piece of A2 paper in half you get two pieces of A3 paper. If you cut a piece of A3 paper in half you get two pieces of A4 paper, and so on.
> For each size, the length and the width are in direct proportion.
>
> A piece of A0 paper has an area of 1 m². Calculate
>
> **a** the constant of proportionality between length and width
>
> **b** the dimensions of a piece of A0 paper, correct to 3 significant figures
>
> **c** the dimensions of a piece of A2 paper, correct to 3 significant figures.
>
>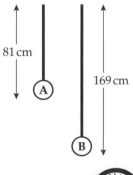

3.2   Other types of proportionality

Links to:

Section 7.7, pp. 120–121

1 P is directly proportional to Q^2. When $Q = 8$, $P = 51.2$.

  a Find a formula for P in terms of Q.

  b Work out P when $Q = 10$.

  c Calculate Q when $P = 180$.

A

2 The mass of copper in grams, m, needed to make a ball bearing is proportional to the cube of its radius in cm, r. When $r = 0.4$ then $m = 2.4$.

  a Find a formula for m in terms of r.

  b Calculate m when $r = 0.6$.

  c Calculate the radius of a ball bearing with mass 218.7 g.

3 The time taken for a clock pendulum to swing once is proportional to the square root of its length. Pendulum A takes 1.8 seconds to swing once.

  a Calculate the time taken for pendulum B to swing once.

  b A clock-maker wants a pendulum that takes 1 second to swing. How long should he make his pendulum?

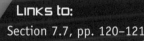

4 Wheelie-bins come in different sizes.
The capacity of the wheelie-bin is proportional to the cube of its height.
A 120 litre wheelie-bin is 0.8 m tall. Calculate

  a the capacity of a wheelie-bin that is 1.5 m tall

  b the height of a 200 litre wheelie-bin.

Give your answers correct to 3 significant figures.

5 Carla wrote this statement in her textbook.

If $a \propto b$ and $b \propto c^2$ then $a \propto c^3$.

Do you agree with Carla's statement?
Give a reason for your answer.

6 y is proportional to x^2 with constant of proportionality k. Which of the following statements is true?
Give a reason for your answer.

   A x is proportional to y^2 with constant of proportionality k

   B x is proportional to y^2 with constant of proportionality $\dfrac{1}{k}$

   C x is proportional to $\sqrt{y}$ with constant of proportionality $\sqrt{k}$

   D x is proportional to $\sqrt{y}$ with constant of proportionality $\dfrac{1}{\sqrt{k}}$

A*

AO3

Challenge yourself

You won't encounter questions like this in the exam, but the underlying maths is covered in your GCSE course. **Have a go!**

Emma is investigating projectiles in physics. She fires a tennis ball into the air and uses a video camera to record its height. She wants to investigate the relationship between height, h m, and time elapsed, t seconds. She records these results.

Time (s)	1	2	3	4
Height (m)	20.1	30.4	30.9	21.6

Emma says that the relationship between t and h can be written as a formula
$$h = kt + jt^2$$
where k and j are constants.

Do you agree with Emma's statement? Give reasons for your answer.

3.3   Inverse proportion

Links to:
Section 7.8, pp. 121–123

1 j is inversely proportional to k. When $k = 4$, $j = 5$.

   a Write a formula for j in terms of k.

   b Calculate k when $j = 8$.

   c Calculate j when $k = 0.2$.

A

2 Two variables are connected by the relationship $s \propto \dfrac{1}{\sqrt{t}}$.

Copy and complete this table of values for s and t.

s	0.8		1.2	0.96
t	9	64		

3 A class is making poppies for Remembrance Day. They need to make 1000 poppies.
The time taken, t minutes, is inversely proportional to the number of people making poppies, n.
With 20 people working, they can make all the poppies in 45 minutes.

   a Write a formula for t in terms of n.

   b How long will it take to make all the poppies if there are only 9 people working?
      Give your answer in hours and minutes.

   c How many people will be needed if the class wants to finish making the poppies in half an hour?

4 This cuboid has a volume of 2160 cm³.

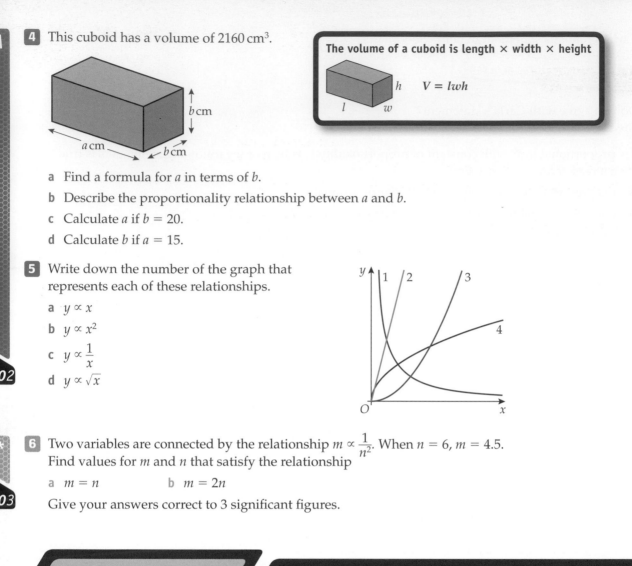

The volume of a cuboid is length × width × height

$V = lwh$

a Find a formula for a in terms of b.

b Describe the proportionality relationship between a and b.

c Calculate a if $b = 20$.

d Calculate b if $a = 15$.

5 Write down the number of the graph that represents each of these relationships.

a $y \propto x$

b $y \propto x^2$

c $y \propto \dfrac{1}{x}$

d $y \propto \sqrt{x}$

6 Two variables are connected by the relationship $m \propto \dfrac{1}{n^2}$. When $n = 6$, $m = 4.5$. Find values for m and n that satisfy the relationship

a $m = n$ b $m = 2n$

Give your answers correct to 3 significant figures.

Challenge yourself

You won't encounter questions like this in the exam, but the underlying maths is covered in your GCSE course. **Have a go!**

An object orbiting at a radius of r km from the centre of the Earth travels with speed v m/s. It takes T minutes to complete one orbit.

$$T \propto \frac{r}{v} \quad \text{and} \quad v \propto \frac{1}{\sqrt{r}}$$

The diagram shows the values of T and r for the Hubble space telescope.

a Find a formula for T in terms of r.

> **Start by writing the proportionality relationships in the question as formulae. Remember to use different letters for the different constants.**

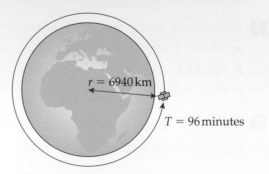

$r = 6940$ km

$T = 96$ minutes

b A geostationary orbit is one that takes exactly one day. An object in a geostationary orbit remains above a fixed position on the Earth's surface. Calculate the radius of a geostationary orbit. Give your answer in km correct to 3 significant figures.

Key Points

Accuracy in calculations   **A**

To find the **greatest possible value** of a calculation you need to use

- the upper bound of any values you add or multiply by
- the lower bound of any values you subtract or divide by.

To find the **least possible value** of a calculation you need to use

- the lower bound of any values you add or multiply by
- the upper bound of any values you subtract or divide by.

Solving problems involving accuracy   **A** **A***

When values have been rounded you sometimes need to choose the upper or the lower bound for a calculation.

Absolute error and percentage error   **A** **A***

The **nominal value** is the value that a quantity is supposed to be if there were no errors.

The **absolute error** is the difference between the measured value and the nominal value.

The **percentage error** is the absolute error written as a percentage of the nominal value.

$$\text{Percentage error} = \frac{\text{Absolute error}}{\text{Nominal value}} \times 100\%$$

4.1   Accuracy in calculations

Links to:
Section 8.4, p.137–140

1 These lengths are measured correct to 1 decimal place.

     $r = 4.0\,\text{m}$ $s = 2.7\,\text{m}$ $t = 10.5\,\text{m}$

Work out the maximum possible value of

a $t - s$ **b** $\dfrac{r}{s}$ **c** $\dfrac{r^2 + s^2}{t}$ **d** $t - (s + r)$ **A**

2 Lizzie has a 4.2 m roll of wrapping paper. She cuts off 0.8 m. Both measurements are correct to 1 decimal place. Calculate the maximum possible length of the remaining wrapping paper.

3 This diagram shows the dimensions of the Statue of Liberty in New York. All the measurements are correct to the nearest metre.   **A**

Work out the maximum possible value of

a the height of the whole statue from A to D

b the distance BC from the head of the statue to its feet

c the distance CD from the top of the head to the top of the torch

d the ratio of the height of the statue BD to the height of the plinth AB.

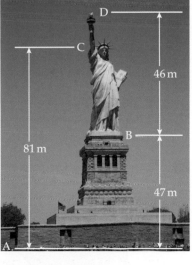

4 The total area of the world's oceans is $3.6 \times 10^8\,\text{km}^2$. The total area of the Earth's surface is $5.1 \times 10^8\,\text{km}^2$. Both measurements are correct to 2 significant figures. Calculate

a the greatest possible value of the area of the world's land masses

b the least possible percentage of the Earth's surface that is covered by ocean.   **AO2**

5 The values given in this newspaper article are correct to 2 significant figures.
Calculate the maximum and minimum values for the mean amount raised for charity by each runner.

8500 RUN IN HALF MARATHON
£1.2 MILLION RAISED FOR CHARITY

6 a Rearrange the formula $A = \dfrac{BC}{C + 1}$ to make C the subject.

b If $A = 0.42$ and $B = 0.30$ (both correct to 2 significant figures) calculate the greatest and least possible values for C.

Challenge yourself

You won't encounter questions like this in the exam, but the underlying maths is covered in your GCSE course. **Have a go!**

Jonah is using the formula $x = \dfrac{n + 20}{n}$ in an investigation.
He records the value of n as 240 to 2 significant figures.
a Use the upper bound for n to calculate a value for x.
b Use the lower bound for n to calculate a value for x.
c Which value of n should Jonah use if he wants to find the greatest possible value of x?
d Prove that your answer to part **c** will always give the greatest possible value of x.

4.2   Problem solving

Links to:
Section 8.4, pp. 137–140

1 Christina needs 1.8 kg of flour for a recipe.
She has two 500 g packets and three 250 g packets.
All the measurements are correct to 2 significant figures.
Does Christina definitely have enough flour for her recipe? Give a reason for your answer.

2 A glass has a capacity of 200 ml.
Karl adds 15 ml of caramel syrup, 35 ml of chocolate syrup and 140 ml of milk to the glass.
All the measurements are correct to 2 significant figures.
Is there any possibility that the glass will overflow? Give a reason for your answer.

3 The dimensions of this cuboid are given correct to the nearest mm.

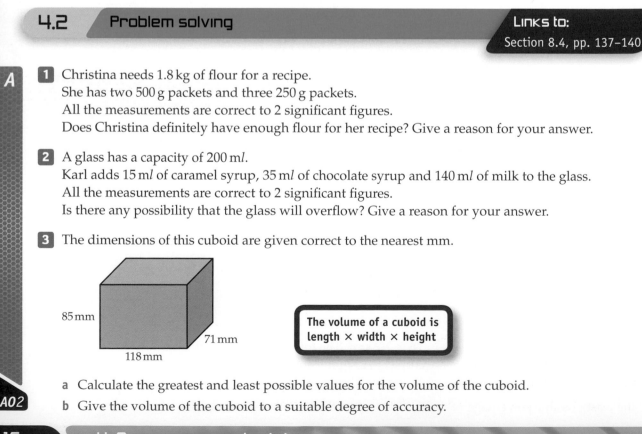

85 mm

71 mm

118 mm

> **The volume of a cuboid is**
> **length × width × height**

a Calculate the greatest and least possible values for the volume of the cuboid.

b Give the volume of the cuboid to a suitable degree of accuracy.

4 A service lift has a maximum loading of 2300 kg, correct to 2 significant figures. How many boxes weighing 60 kg, correct to 1 significant figure, can safely be loaded onto the lift?
Give a reason for your answer.

5 Uchenna is painting his bedroom. The total area he needs to paint is 55 m², to the nearest whole number. 1 litre of paint will cover an area of 13 m².
Uchenna has 9 litres of paint, correct to the nearest litre.
Will he definitely have enough paint to cover his bedroom with two complete coats?
Show working to support your answer.

6 A fairground game requires the player to throw a circular hoop over a square wooden block to win a prize.
The hoop has a radius of 11.7 cm (to 3 significant figures) and the wooden block has a side length of 16 cm (to the nearest cm).
Will the hoop definitely fit over the wooden block?
Show working to support your answer.

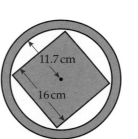

AO3

Challenge yourself

You won't encounter questions like this in the exam, but the underlying maths is covered in your GCSE course. **Have a go!**

The diagram shows a children's game. A wooden cube must be passed through a hole in the shape of a regular octagon. The measurements are correct to the nearest cm.

Show that the cube will definitely be able to pass through the hole.

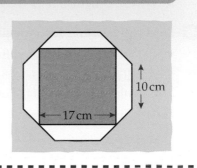

4.3   Percentage error

Links to:
Section 8.5, pp. 141–143

1 A chocolate bar labelled 400 g was found to weigh 403.5 g.
Calculate
  a the absolute error   **b** the percentage error.

A

2 The measurements on this cuboid are correct to 1 decimal place.
Calculate
  a the nominal value for the volume of the cuboid
  b the maximum absolute error for the volume of the cuboid
  c the maximum percentage error for the volume of the cuboid.

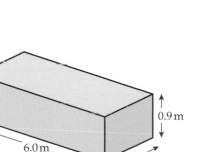

3 A passenger plane travels for 3 minutes at a speed of 245 m/s.
Both measurements are correct to the nearest whole number.
  a Calculate the nominal distance travelled by the plane in km.
  b Calculate the maximum absolute error for the distance travelled.
  c Calculate the maximum percentage error for the distance travelled.

A

4 This diagram shows a side view of the Millau suspension bridge in France.
All the measurements are correct to the nearest metre.

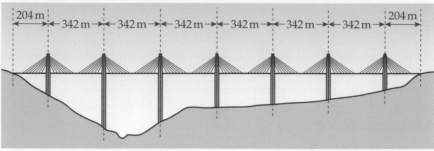

AO2

a Calculate the maximum possible percentage error for the length of one 342 m section of the bridge.

b Calculate the maximum possible percentage error for the length of the whole bridge.

A*

5 Two machines produce different components at a factory. The factory manager selects a sample of three components from each machine and measures their lengths.

	Nominal component length	Sample 1	Sample 2	Sample 3
Machine A	2216 mm	2210 mm	2224 mm	2221 mm
Machine B	814 mm	810 mm	822 mm	809 mm

a Calculate the mean percentage error for each machine, correct to 3 significant figures.

AO3

b Write a conclusion to say which machine is performing more accurately.

A*

6 Steel ball bearings for a car should be made using 14.1 cm^3 of steel.

a Calculate the nominal value for the diameter of one ball bearing.

b A ball bearing is made using 13.6 cm^3 of steel.
Calculate the percentage error in the diameter of the ball bearing.

AO2

Give your answers correct to 3 significant figures.

> The volume of a sphere of radius r is $\frac{4}{3}\pi r^3$

Challenge yourself

You won't encounter questions like this in the exam, but the underlying maths is covered in your GCSE course. **Have a go!**

Quantity p has a maximum percentage error of $X\%$.
Quantity q has a maximum percentage error of $Y\%$.
Investigate the maximum percentage error of the quantity pq.
Find an expression for the maximum percentage error of pq in terms of X and Y.

> You could begin by experimenting with different values for p, q, X and Y.

Links to:
Higher Student Book
Ch13 pp. 200–201,
Ch14 pp. 210–213

Key Points

Recurring decimals   **A**

Every recurring decimal can be written as a fraction.
To convert a recurring decimal to a fraction:

- Multiply the number by a power of ten.

- Subtract the original decimal from the result.
  This leaves you with only whole numbers.

- Solve the equation.

$n = 0.7\dot{2}$

> If there are n recurring digits you need to multiply the decimal by 10^n.

$100n = 72.7\dot{2}$

$99n = 72$

$n = \frac{72}{99} = \frac{8}{11}$

Simultaneous equations   **A**

Simultaneous equations are pairs of equations which have two unknowns. You need to find values of the unknowns which make both equations true at the same time.
You can solve simultaneous equations by eliminating one of the unknowns.

①　　　　$3x - 2y = 17$

②　　　　$8x + 3y = 12$

$3 \times$ **①**　　$9x - 6y = 51$

$2 \times$ **②**　　$\underline{16x + 6y = 24}$

　　　　$25x \qquad = 75$

　　　　　　$x = 3$

> Add these two equations to eliminate y.

Substitute into **①**

　　　$9 - 2y = 17$

　　　　　$y = -4$

> Multiply each equation by a number until two of the coefficients are the same. Then add or subtract the equations to eliminate one of the unknowns.

5.1   Converting recurring decimals to fractions

Links to:
Section 13.4, pp. 200–201

1 Convert each recurring decimal into a fraction in its simplest form.   **A**

　　a $0.\dot{5}$　　　b $0.3\dot{6}$　　　c $0.9\dot{5}$　　　d $0.\dot{2}\dot{4}$　　　e $0.10\dot{9}$　　　f $0.05\dot{2}$

2 Convert each recurring decimal into a fraction in its simplest form.

　　a $0.30\dot{6}$　　b $0.\dot{2}6\dot{7}$　　c $0.00\dot{1}$　　d $0.000\dot{1}$　　e $0.56\dot{1}\dot{9}$　　f $0.\dot{3}30\dot{3}$

3 Write $4.5\dot{6}$ in the form $\frac{a}{b}$, where a and b are whole numbers.   **A** **AO2**

4 Work out $0.3\dot{6} + 0.\dot{2}\dot{1}$. Write your answer as a fraction in its lowest terms.   **A**

5 Show that $0.\dot{2}8571\dot{4}$ is equal to $\frac{2}{7}$.   **AO3**

6 A bag of sweets weighs exactly 2 kg. Each sweet weighs $14.\dot{8}1\dot{4}$g. How many sweets are in the bag?   **A*** **AO3**

Challenge yourself

You won't encounter questions like this in the exam, but the underlying maths is covered in your GCSE course. **Have a go!**

Use working to show that the infinite sum $\frac{1}{2} + \frac{1}{20} + \frac{1}{200} + \frac{1}{2000} + \dots$ is equal to $\frac{5}{9}$.

A

1 This rectangle is made from 7 identical smaller rectangles. Find the area of the large rectangle.

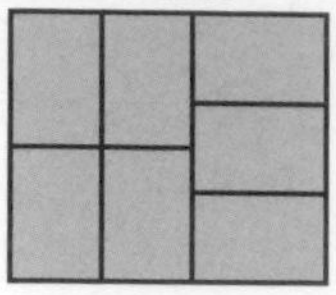

←—— 14 cm ——→

2 Will has written a fraction $\frac{a}{b}$.
If he adds 4 to the numerator and the denominator, the new fraction is equivalent to $\frac{2}{3}$.
If he subtracts 1 from the numerator and the denominator, the new fraction is equivalent to $\frac{1}{2}$.
Find the values of a and b.

3 Gavin is conducting a chemistry experiment.
He has a 15% sulphuric acid solution and a 23% sulphuric acid solution.
He needs to combine these to create a 20% sulphuric acid solution.
How much of each solution should he use to create 100 ml of 20%
sulphuric acid solution?

> **A 15% sulphuric acid solution contains 15% sulphuric acid and 85% distilled water.**

4 The moving walkway at an airport is 60 m long.
It takes Nick 24 seconds to walk the length of the walkway in one
direction, and 1 minute and 20 seconds in the other.
Calculate the speed of the moving walkway.

> **You will need to use the formula speed $= \dfrac{\text{distance}}{\text{time}}$**

AO2

A

5 A paint manufacturer makes three different shades of green.

APPLE WHITE
RATIO
WHITE : GREEN
4:1

HINT OF MINT
RATIO
WHITE : GREEN
7:3

SUMMER FIELDS
RATIO
WHITE : GREEN
1:1

Hamish has run out of *Hint of Mint*, and wants to mix *Apple White* with *Summer Fields* to create more.
How much of each type of paint should Hamish use to mix 15 litres of *Hint of Mint*?

6 In a magic square the sum of the numbers in any row,
column or long diagonal is the same.
The diagram shows a magic square.
Work out the values of a, b, c and d.

a	b	12
c	d	7
6	13	8

AO3

Challenge yourself

You won't encounter questions like this in the exam, but the
underlying maths is covered in your GCSE course. **Have a go!**

Solve these simultaneous equations.
$$7a + 2b + 2c = 35$$
$$11a + 5b + 2c = 54$$
$$-3a + b - 4c = -21$$

> **You need to find the values of a, b and c which make all three equations true.**

6 Indices and surds

Links to:
Higher Student Book
Ch16, pp. 226–232, 235–240

Key Points

Powers and roots A A*

You can use the rules of indices to simplify expressions involving powers and roots.

- $x^m \times x^n = x^{m+n}$
- $\dfrac{x^m}{x^n} = x^{m-n}$
- $(x^m)^n = x^{mn}$
- $x^1 = x$
- $x^0 = 1$
- $x^{-n} = \dfrac{1}{x^n}$
- $\sqrt[n]{x} = x^{\frac{1}{n}}$
- $\sqrt[n]{x^m} = \left(\sqrt[n]{x}\right)^m = x^{\frac{m}{n}}$

Surds A A*

Numbers that cannot be written as a fraction are called **irrational numbers**. Square roots of whole numbers that are not perfect squares are irrational. You can write these numbers exactly by using surds.

You can simplify expressions involving surds using these rules:

- $\sqrt{ab} = \sqrt{a}\,\sqrt{b}$
- $\sqrt{\dfrac{a}{b}} = \dfrac{\sqrt{a}}{\sqrt{b}}$

> $\sqrt{10}$ and $3 - 2\sqrt{2}$ are examples of expressions using surds.

If a fraction has a surd on the bottom it is sometimes useful to **rationalise the denominator.**

$$\frac{5}{2\sqrt{3}} = \frac{5}{2\sqrt{3}} \times \frac{\sqrt{3}}{\sqrt{3}} = \frac{5\sqrt{3}}{2\sqrt{3} \times \sqrt{3}} = \frac{5\sqrt{3}}{6}$$

6.1 Simplifying powers and roots

Links to:
Section 16.1, pp. 226–229

1 Write each of these expressions as a single power.

a $(8^2)^2$ b $(10^3)^5$ c $(2^8)^3$ d $(4^{-1})^6$ e $(9^2)^{-5}$ f $(6^{-4})^{-4}$

2 Write each of these expressions as a single power.

a $\sqrt{2^4}$ b $\sqrt{8^{10}}$ c $\sqrt[3]{14^9}$ d $\sqrt[6]{2^{24}}$ e $\sqrt[3]{5^{12}}$ f $\sqrt[10]{50^{50}}$

3 Work out the value of

a $\sqrt{5^4}$ b $\sqrt{3^6}$ c $(2^2)^3$ d $\sqrt[3]{6^9}$ e $(4^{-2})^{-2}$ f $\sqrt[8]{7^{24}}$

4 Find the value of n in each equation.

a $\dfrac{2 \times \sqrt{4^2}}{2^2} = 2^n$ b $\dfrac{7^9}{49^2} = 7^n$ c $\dfrac{3^5 \times \sqrt{3^{16}}}{(3^2)^3} = 3^n$ d $\dfrac{(2^4)^5 \times \sqrt[4]{16^3}}{4^2} = 2^n$

5 Work out the number of digits in $\sqrt[10]{100^{90}}$

6 A cube has a volume of $3^{12}\,\text{cm}^3$. Work out the length of one side.

7 Find values of x and y that satisfy the equation $\dfrac{9 \times (6^2)^5}{4^2 \times \sqrt{3^4}} = 3^x \times \sqrt[y]{2^{24}}$

> You need to consider the powers of 2 and the powers of 3 separately.

A

A

AO2

A*

AO3

Challenge yourself

You won't encounter questions like this in the exam, but the underlying maths is covered in your GCSE course. **Have a go!**

Solve the simultaneous equations

$$4^a \times 8^b = 16$$
$$\frac{5^a}{25^b} = (5^3)^3$$

Links to:
Section 16.4, pp. 229–232

A

1 Work out the value of

a $49^{\frac{1}{2}}$ b $121^{0.5}$ c $81^{\frac{1}{4}}$ d $125^{\frac{1}{3}} \times 3^{-2}$ e $100^{\frac{1}{2}} \times 5^{-2}$ f $32^{\frac{1}{5}} \times 9^{\frac{1}{2}}$

2 Write each of these as a power of 2.

a $\sqrt[5]{2^3}$ b $\sqrt[3]{4}$ c $\sqrt{\sqrt{2}}$ d $\sqrt[3]{16}$ e $\sqrt[3]{8^2}$ f $\sqrt[3]{\sqrt[3]{32}}$

3 Work out the value of

a $216^{\frac{2}{3}}$ b $9^{1.5}$ c 7^1 d $81^{\frac{3}{4}}$ e $4^{\frac{5}{2}}$ f $(6^4)^0$

A
AO3

4 Solve the equation $x^{(2x+6)} = 1$.

A*

5 Work out the value of

a $100^{-\frac{1}{2}}$ b $125^{-\frac{2}{3}}$ c $\left(\frac{4}{9}\right)^{-\frac{1}{2}}$ d $\left(\frac{16}{81}\right)^{-\frac{3}{4}}$

A*
AO2

6 Write these numbers in order of size, smallest first.

5^{-2}, 5^0, $125^{-\frac{1}{2}}$, $\sqrt[3]{5}$, $\sqrt{\frac{1}{5^2}}$

Write each number as a power of 5.

A*
AO3

7 Evaluate $16^{(-2)^{-2}}$.

Challenge yourself

You won't encounter questions like this in the exam, but the underlying maths is covered in your GCSE course. **Have a go!**

Solve this equation to find the value of n.

$$\left(\frac{81}{64}\right)^n \times \left(\frac{16}{9}\right)^{-\frac{3}{2}} = \sqrt{\frac{9}{64}}$$

6.3 Surds

Links to:
Section 16.4, pp. 235–237

A

1 Evaluate these expressions.

a $\sqrt{24} \times \sqrt{6}$

b $\sqrt{5} \times \sqrt{20}$

c $6\sqrt{5} \times 10\sqrt{5}$

d $2\sqrt{3} \times 5\sqrt{12}$

e $\frac{\sqrt{6}}{2} \times \frac{\sqrt{24}}{3}$

f $\frac{\sqrt{45}}{3} \times 2\sqrt{5}$

2 Evaluate these expressions.

a $\frac{\sqrt{28}}{\sqrt{7}}$

b $\sqrt{75} \div \sqrt{3}$

c $\frac{3\sqrt{108}}{2\sqrt{3}}$

d $\sqrt{32} \div 2\sqrt{2}$

e $\frac{\sqrt{45}}{2\sqrt{5}}$

f $5\sqrt{180} \div 6\sqrt{5}$

3 Write each expression in the form $\sqrt{a}$ where a is a whole number.

a $\sqrt{2} \times \sqrt{3}$

b $5\sqrt{7}$

c $3\sqrt{2} \times \sqrt{5}$

d $2\sqrt{7} \times 3\sqrt{11}$

e $\frac{12}{\sqrt{3}}$

f $\frac{5}{\sqrt{7}} \times \frac{7}{\sqrt{5}}$

4 Calculate the unknown lengths in these triangles. Leave your answers in surd form.

a

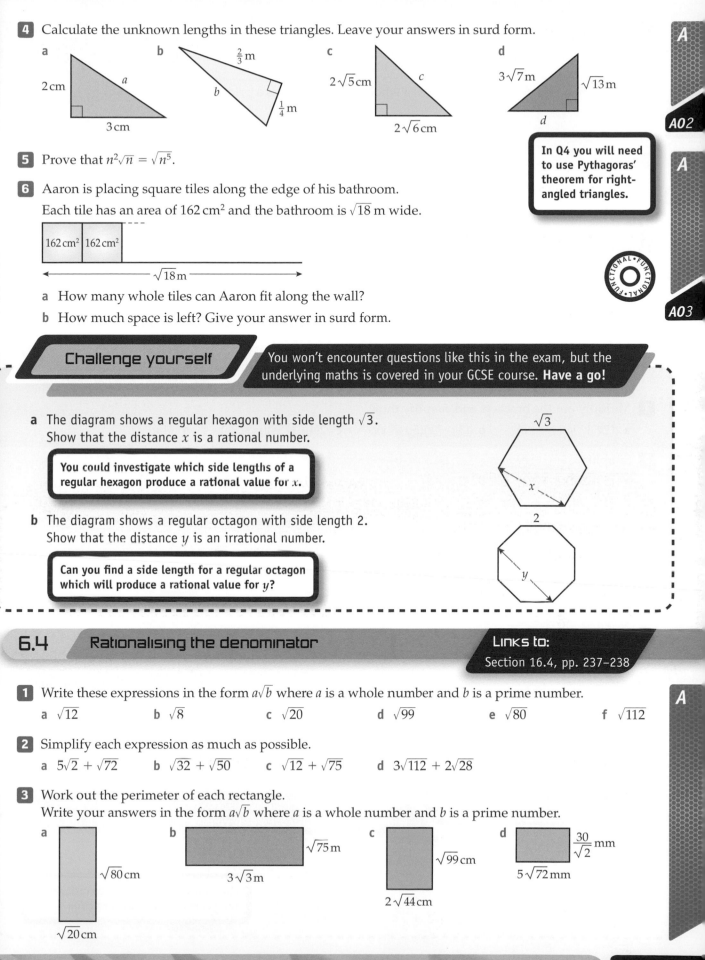

2 cm a 3 cm

b

$\frac{2}{3}$ m b $\frac{1}{4}$ m

c

$2\sqrt{5}$ cm c $2\sqrt{6}$ cm

d

$3\sqrt{7}$ m $\sqrt{13}$ m d

> **In Q4 you will need to use Pythagoras' theorem for right-angled triangles.**

5 Prove that $n^2\sqrt{n} = \sqrt{n^5}$.

6 Aaron is placing square tiles along the edge of his bathroom.
Each tile has an area of $162\ cm^2$ and the bathroom is $\sqrt{18}$ m wide.

162 cm² | 162 cm²

$\sqrt{18}$ m

a How many whole tiles can Aaron fit along the wall?

b How much space is left? Give your answer in surd form.

Challenge yourself

You won't encounter questions like this in the exam, but the underlying maths is covered in your GCSE course. **Have a go!**

a The diagram shows a regular hexagon with side length $\sqrt{3}$.
Show that the distance x is a rational number.

> You could investigate which side lengths of a regular hexagon produce a rational value for x.

$\sqrt{3}$

x

b The diagram shows a regular octagon with side length 2.
Show that the distance y is an irrational number.

> Can you find a side length for a regular octagon which will produce a rational value for y?

2

y

6.4   Rationalising the denominator

Links to:
Section 16.4, pp. 237–238

1 Write these expressions in the form $a\sqrt{b}$ where a is a whole number and b is a prime number.

a $\sqrt{12}$ b $\sqrt{8}$ c $\sqrt{20}$ d $\sqrt{99}$ e $\sqrt{80}$ f $\sqrt{112}$

2 Simplify each expression as much as possible.

a $5\sqrt{2} + \sqrt{72}$ b $\sqrt{32} + \sqrt{50}$ c $\sqrt{12} + \sqrt{75}$ d $3\sqrt{112} + 2\sqrt{28}$

3 Work out the perimeter of each rectangle.
Write your answers in the form $a\sqrt{b}$ where a is a whole number and b is a prime number.

a $\sqrt{80}$ cm $\sqrt{20}$ cm

b $\sqrt{75}$ m $3\sqrt{3}$ m

c $\sqrt{99}$ cm $2\sqrt{44}$ cm

d $\frac{30}{\sqrt{2}}$ mm $5\sqrt{72}$ mm

A

AO2

A

AO3

A

4 Simplify each expression by rationalising the denominator.

a $\dfrac{2}{\sqrt{2}}$ b $\dfrac{4}{5\sqrt{3}}$ c $\dfrac{2}{10\sqrt{6}}$ d $\dfrac{2\sqrt{2}}{\sqrt{7}}$ e $\dfrac{4\sqrt{12}}{\sqrt{5}}$ f $\dfrac{12\sqrt{5}}{5\sqrt{3}}$

5 Write $\sqrt[3]{56}$ in the form $a\sqrt[3]{b}$ where a is a whole number and b is a prime number.

6 Write $\dfrac{1}{10\sqrt{3}} + \dfrac{4}{5\sqrt{12}}$ in the form $a\sqrt{b}$ where a is a rational number and b is a prime number.

> **A rational number can be written as a fraction with whole numbers.**

Challenge yourself

You won't encounter questions like this in the exam, but the underlying maths is covered in your GCSE course. **Have a go!**

Write $\dfrac{2}{2 + \sqrt{3}}$ in the form $a + b\sqrt{3}$ where a and b are whole numbers.

6.5   More complicated surds

> **Links to:**
> Section 16.4, pp. 239–240

1 Multiply out the brackets and simplify these.

a $(2a + 3b)(4a - b)$ b $(3y - x)(5y + x)$ c $2(4p + 5q)(p - 2q)$ d $(4m - n)^2$

2 Simplify

a $(5 + \sqrt{2})(3 + \sqrt{2})$ b $(2 + \sqrt{8})^2$ c $(1 + \sqrt{7})(2 - \sqrt{7})$

d $(4 + \sqrt{18})(5 + \sqrt{50})$ e $(3 - \sqrt{40})(10 - \sqrt{10})$ f $(7 - \sqrt{5})^2$

3 Write each expression as a whole number.

a $(8 - \sqrt{32})(2 + \sqrt{2})$ b $(5 + \sqrt{75})(1 - \sqrt{3})$ c $(6 + \sqrt{18})(12 - \sqrt{72})$ d $(3 - \sqrt{7})(6 + \sqrt{28})$

4 Write each expression as a whole number.

a $(5 + \sqrt{5})(5 - \sqrt{5})$ b $(10 + \sqrt{8})(10 - \sqrt{8})$ c $(6 + \sqrt{14})(6 - \sqrt{14})$ d $(12 + \sqrt{20})(12 - \sqrt{20})$

5 Solve the equation $(9 + \sqrt{5})x = 76$.

> **Look at your answers to Q4.**

6 Work out the volume of a cube with side length $(2 + \sqrt{3})$ cm.
Give your answer in the form $a + b\sqrt{3}$, where a and b are whole numbers.

7 Multiply out $(2\sqrt{3} + \sqrt{2})(5\sqrt{3} - 6\sqrt{2})$.
Give your answer in the form $a + b\sqrt{c}$ where a, b and c are whole numbers.

Challenge yourself

You won't encounter questions like this in the exam, but the underlying maths is covered in your GCSE course. **Have a go!**

a Simplify $(\sqrt{a} + \sqrt{b})(\sqrt{a} - \sqrt{b})$.

b Find the value of the sum

$$\dfrac{1}{\sqrt{1} + \sqrt{2}} + \dfrac{1}{\sqrt{2} + \sqrt{3}} + \dfrac{1}{\sqrt{3} + \sqrt{4}} + \ldots + \dfrac{1}{\sqrt{24} + \sqrt{25}}$$

> **Use your answer to part a to rationalise the denominator of each fraction.**

Links to:
Higher student book
Ch17, pp. 254–255
Ch19, pp. 283–284, 291–293

Key Points

Proof   A   A*

You can use algebra to prove that something is true. You need to write down each step of your working so somebody else can understand your proof.

You can use these facts to help you prove number statements:

- Even numbers can be written in the form $2n$ where n is a whole number.
- Odd numbers can be written in the form $2n + 1$ where n is a whole number.
- Multiples of 3 can be written in the form $3n$ where n is a whole number.

You can show that a statement is not true by providing a counter-example.

Perpendicular lines   A

Perpendicular lines are straight lines that cross each other at right angles.

If one line has gradient m, the other line has gradient $-\dfrac{1}{m}$.

The product of the gradients of two perpendicular lines is always -1.

> **Straight-line graphs have equations of the form**
> $y = mx + c$.
> m **is the gradient and** c **is the** y**-intercept.**

Velocity-time graphs   A

The gradient of a **velocity-time** graph represents the acceleration of an object.

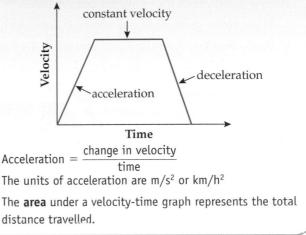

$$\text{Acceleration} = \frac{\text{change in velocity}}{\text{time}}$$

The units of acceleration are m/s² or km/h²

The **area** under a velocity-time graph represents the total distance travelled.

7.1   Proof

Links to:
Section 17.6, pp. 254–255

1 The diagram shows an addition pyramid.   *A*

```
            34
        16      18
      9     7      11
    4     5     2     9
```

Each number is the sum of the two numbers below it.

Prove that if the base numbers of the pyramid are all odd, the top number will be even.

2   **a** Prove this statement:

      The difference of the squares of two consecutive even numbers is always divisible by 4.

   b Is this statement also true for odd numbers? Give a reason for your answer.

3   **a** Use a counter-example to show that this statement is not true for all numbers a and b:

      $a^2 + b^2 < (a + b)^2$

   b Prove that the statement is true as long as a and b are positive.   **AO2**

4 n is a positive whole number.   *A*

   a Explain why $n(n + 1)$ must be an even number.

   b Prove that the sum of the squares of two consecutive whole numbers is 1 more than a multiple of 4.   **AO3**

5 Jonah writes down the three-digit number 423. He adds together the digits to get $4 + 2 + 3 = 9$. Jonah says that this means 423 is divisible by 9.

a Confirm that 423 is divisible by 9.

b Prove that Jonah's method will always work.

> You can write a three-digit number '*ABC*' as $100A + 10B + C$.

6 Alison has worked out a shortcut for squaring two-digit numbers that end in 5. She has tried her method with the number 75.

a Confirm that $75^2 = 5625$.

b Check Alison's method works for the number 15.

c Prove that Alison's method will work for all two-digit numbers that end in 5.

> Square the first digit. $7^2 = 49$
> Add the first digit to the result. $49 + 7 = 56$
> Write 25 at the end.   5625

7 A **palindrome** is a number that is the same written forwards or backwards. All the numbers in the cloud are palindromes.

a How many three-digit palindromes are there?

b Prove that any four-digit palindrome is divisible by 11.

> 99
> 363
> 4004
> 12 321

Challenge yourself

You won't encounter questions like this in the exam, but the underlying maths is covered in your GCSE course. **Have a go!**

Leo says that the relationship $F + V - E = 2$ is true for all solid shapes, where F = number of faces, V = number of vertices and E = number of edges. The diagram shows a prism with a hexagonal cross-section.

a Confirm that $F + V - E = 2$ for this hexagonal prism.

b Prove that $F + V - E = 2$ for any prism whose cross-section is an n-sided polygon.

c Prove Leo's theorem for a pyramid whose base is an n-sided polygon.

> Investigate Leo's theorem for other solid shapes.

7.2   Perpendicular lines

Links to: Section 19.3, pp. 283–284

1 Write down the gradient of a line that is perpendicular to each of these lines.

a $y = 2x + 5$ b $y = 5x - 2$ c $y = -\frac{1}{3}x + 1$

d $y = -3x + 6$ e $y = \frac{2}{3}x - 9$ f $y = -\frac{2}{5}x - \frac{1}{3}$

2 Find the equation of the line perpendicular to the line $y = 4x + 6$ which passes through the point $(0, 5)$.

3 Show that the line $2x - 3y + 5 = 0$ is perpendicular to the line $4y + 6x - 1 = 0$.

4 Find the equation of the line perpendicular to the line $2y + 3x = 8$ which passes through the point $(0, -4)$.

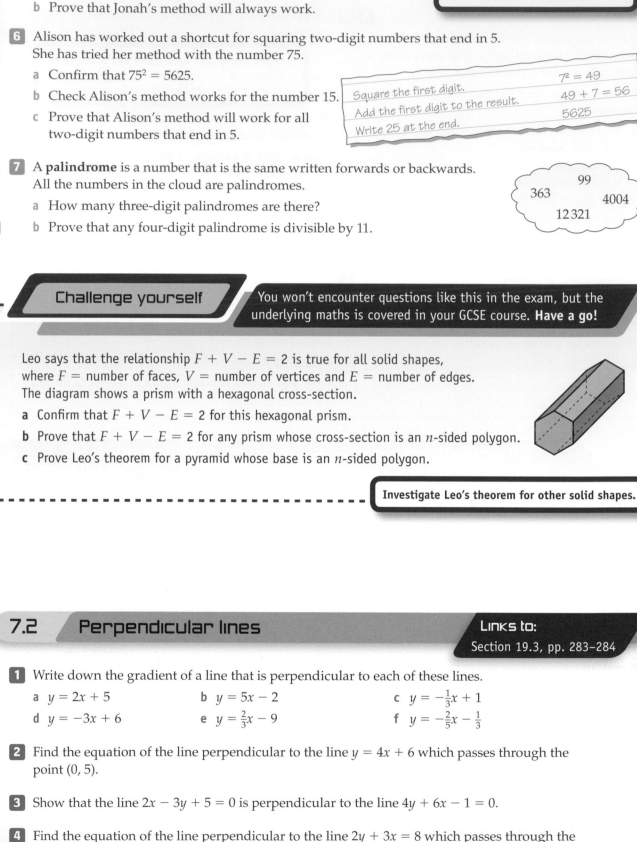

5 The diagram shows a line drawn on a coordinate grid.
Write down the equation of the line that is perpendicular
to this line and passes through the point $(0, -1)$.

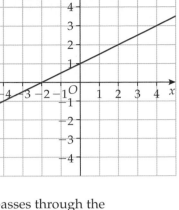

6 Find the equation of the line perpendicular to the line $4y - x = 10$ which passes through the point $(2, -1)$.

7 Find the equation of the perpendicular bisector of the line segment joining the points $(1, 2)$ and $(5, 4)$.

8 Line A passes through the points $(3, 0)$ and $(6, -2)$. Line B passes through the points $(1, 1)$ and $(3, 4)$.
Show that the lines A and B are perpendicular.

9 This diagram shows the edge of a field.

A fence must be constructed so that it is perpendicular to the edge of the field and passes through point P.
Find the equation of the line of the fence.

Challenge yourself

You won't encounter questions like this in the exam, but the underlying maths is covered in your GCSE course. **Have a go!**

A triangle has vertices at $(-1, -2)$, $(3, -1)$ and $(3, 2)$.

a Show that the perpendicular bisectors of the three sides of the triangle meet at a single point.

b Investigate this relationship for other triangles.

A

1 The diagram shows a velocity-time graph for a fairground ride.

Use the diagram to calculate

a   the acceleration for the first 10 seconds of the ride

b   the total distance travelled during the ride.

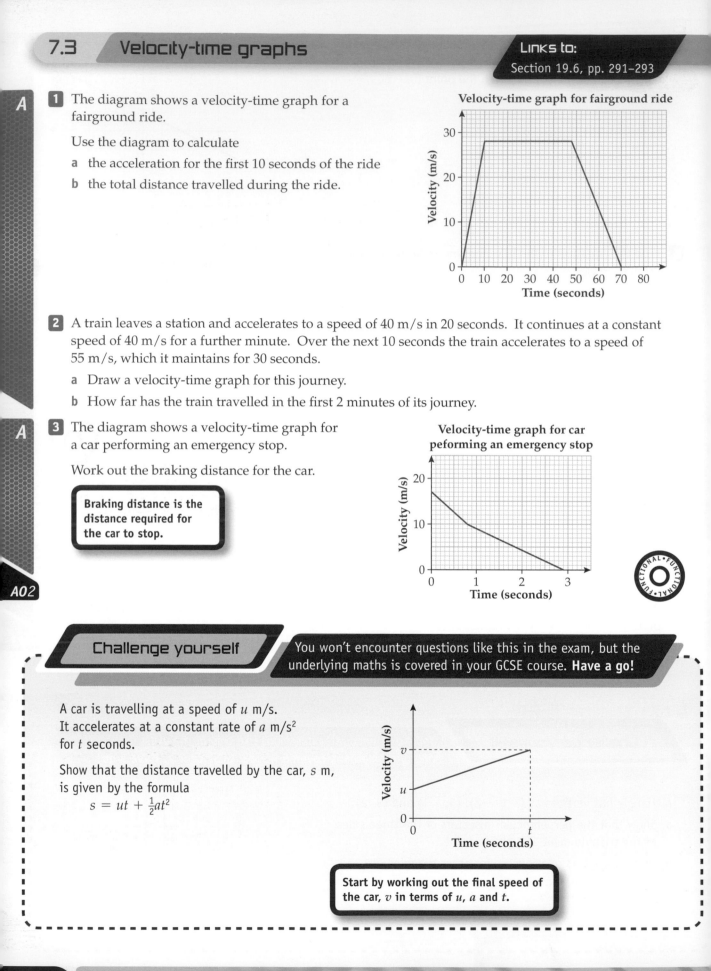

Velocity-time graph for fairground ride

2 A train leaves a station and accelerates to a speed of 40 m/s in 20 seconds. It continues at a constant speed of 40 m/s for a further minute. Over the next 10 seconds the train accelerates to a speed of 55 m/s, which it maintains for 30 seconds.

a   Draw a velocity-time graph for this journey.

b   How far has the train travelled in the first 2 minutes of its journey.

A

3 The diagram shows a velocity-time graph for a car performing an emergency stop.

Work out the braking distance for the car.

Braking distance is the distance required for the car to stop.

Velocity-time graph for car peforming an emergency stop

AO2

Challenge yourself

You won't encounter questions like this in the exam, but the underlying maths is covered in your GCSE course. **Have a go!**

A car is travelling at a speed of u m/s. It accelerates at a constant rate of a m/s^2 for t seconds.

Show that the distance travelled by the car, s m, is given by the formula
$$s = ut + \tfrac{1}{2}at^2$$

Start by working out the final speed of the car, v in terms of u, a and t.

Links to:
Higher Student Book
Ch20, pp. 298–299,
305–314

Key Points

Factorising quadratic expressions   **A**

You can use the following method to factorise a quadratic expression of the form $ax^2 + bx + c$.

- Start with a quadratic expression:
  $4x^2 + 8x + 3$

- Find a factor pair of ac that add up to b:
  2 and 6 are factors of 12 that add up to 8

- Rewrite the x term as the sum of these factors:
  $4x^2 + 2x + 6x + 3$

- Factorise the first and second pairs of terms:
  $2x(2x + 1) + 3(2x + 1)$

- Factorise again:
  $(2x + 3)(2x + 1)$

Solving quadratic equations by factorising   **A**

To solve a quadratic equation, rearrange it so that one side is equal to 0. If the quadratic expression can be factorised, the solutions (or roots) will be the values of the unknown that make either factor equal to 0.

The quadratic formula   **A** **A***

Not all quadratic equations can be factorised. The solutions to the quadratic equation $ax^2 + bx + c = 0$ are given by the formula

$$x = \frac{-b \pm \sqrt{b^2 - 4ac}}{2a}$$

The discriminant   **A***

A quadratic equation can either have two solutions, one solution or no solutions. The discriminant is the expression $b^2 - 4ac$ under the square root in the quadratic formula.

- If $b^2 - 4ac > 0$ there are two solutions.
- If $b^2 - 4ac = 0$ there is one solution.
- If $b^2 - 4ac < 0$ there are no solutions.

> You can't calculate the square root of a negative number, so if the discriminant is negative there are no solutions.

Completing the square   **A***

You can solve a quadratic equation directly by completing the square. You can use this identity to help you complete the square:

$$x^2 + 2bx + c = (x + b)^2 - b^2 + c$$

> Once you have completed the square the unknown only appears once in the equation. You can solve the equation using inverse operations.

8.1   Factorising the difference of two squares

Links to:
Section 20.1, pp. 298–299

1   Factorise these quadratic expressions.

   a $10x^2 - 90$ b $6m^2 - 54$ c $5q^2 - 20$

   d $32a^2 - 18$ e $12 - 27x^2$ f $200 - 98d^2$

> If an expression can be written as the difference of two squares you can factorise it using the identity
> $a^2 - b^2 = (a + b)(a - b)$

2   Factorise

   a $2x^2 - 2y^2$ b $12a^2 - 3b^2$ c $45m^2 - 20n^2$

   d $32q^2 - 50p^2$ e $8y^2 - 50z^2$ f $48s^2 - 147t^2$

3   Factorise $9a^4 - 25b^2c^2$.

A

A02

4 The diagram shows two intersecting squares.

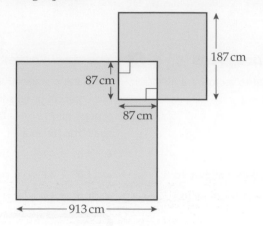

Work out the total shaded area without using a calculator.

Show all your working.

5 Write $16a^4 - 81b^4$ as the product of three factors.

6 Prove that if n is a whole number greater than 2, then $n^4 - 1$ has at least four proper factors.

> **A proper factor is any factor other than n or 1.**

Challenge yourself

You won't encounter questions like this in the exam, but the underlying maths is covered in your GCSE course. **Have a go!**

Prove that there is exactly one way to write any prime number other than 2 as the difference of two squares.

8.2 Factorising quadratics of the form $ax^2 + bx + c$

Links to:
Section 20.4, pp. 305–308

1 Factorise these expressions.

a $2x^2 + 10x + 12$ b $5x^2 + 35x + 60$ c $4x^2 - 12x + 8$

d $10x^2 + 10x - 60$ e $3x^2 - 6x - 24$ f $100x^2 - 100x - 3000$

> **You need to take out a common factor first.**

2 Factorise

a $2x^2 + 5x + 2$ b $3x^2 + 11x + 6$ c $2x^2 - 7x + 6$

d $5x^2 + 4x - 1$ e $3x^2 - 5x - 12$ f $7x^2 + 3x - 4$

3 Factorise

a $4x^2 + 8x + 3$ b $4x^2 + 7x + 3$ c $6x^2 + 5x + 1$

d $6x^2 + 13x + 6$ e $4x^2 - 4x - 15$ f $10x^2 + x - 21$

4 Factorise

a $9x^2 - 6x - 3$ b $15x^2 - 18x + 3$ c $20x^2 + 170x + 150$

d $20x^2 - 60x - 135$ e $\frac{9}{2}x^2 + \frac{3}{2}x - 1$ f $6x^2 + 14x + 7.5$

5 Show that $\dfrac{2x^2 + 7x + 6}{x + 2} = 2x + 3$.

6 A rectangle has an area of $(3n^2 + 17n + 10)\,\text{cm}^2$.
One side has a length of $(n + 5)\,\text{cm}$.
Find an expression for the length of the other side.

$\xleftarrow{\hspace{2cm}} ? \xrightarrow{\hspace{2cm}}$

$\boxed{\text{Area} = (3n^2 + 17n + 10)\,\text{cm}^2}$ $(n + 5)\,\text{cm}$

A02

7 Show that if n is a whole number, then $4n^2 + 12n + 9$ is a square number.

*A

8 Write $4x^4 - 5x^2 - 9$ as the product of three factors.

A03

Challenge yourself

You won't encounter questions like this in the exam, but the underlying maths is covered in your GCSE course. **Have a go!**

a Write $4x^4 - 13x^2 + 9$ as the product of four factors.

b Use your answer to part **a** to solve the equation $4x^4 - 13x^2 + 9 = 0$

8.3 Solving quadratic equations by factorising

Links to: Section 20.4, pp. 308–309

1 Solve

   a $(2x - 1)(x - 10) = 0$
   b $(4x + 3)(x - 9) = 0$
   c $(10x - 7)(x + 3) = 0$

   d $2(x + 1)(5x - 2) = 0$
   e $4(2x + 5)(x - 7) = 0$
   f $10(4x - 11)(2x + 11) = 0$

A

2 Solve

   a $2x^2 - 13x + 15 = 0$
   b $5x^2 + 38x + 21 = 0$
   c $5x^2 + 4x = 1$

   d $x = 2x^2 - 28$
   e $4x^2 + 15x - 4 = 0$
   f $4x^2 + 12x = 27$

3 Solve

   a $6x^2 - 24x + 24 = 0$
   b $12x^2 - 48x = 27$
   c $x(2x - 3) = 5$

   d $(2x + 1)^2 = x + 2$
   e $-18x = (10x + 1)(x - 2)$
   f $(4x + 1)(x - 10) = 5 - 35x$

4 Solve the equation $\quad 5x + 3 = \sqrt{3x + 7}$.

A

5 **a** This shape has an area of $20\,\text{m}^2$. Find the value of x.

   b Supraj says there are two answers to part **a**. Do you agree?
   Give a reason for your answer.

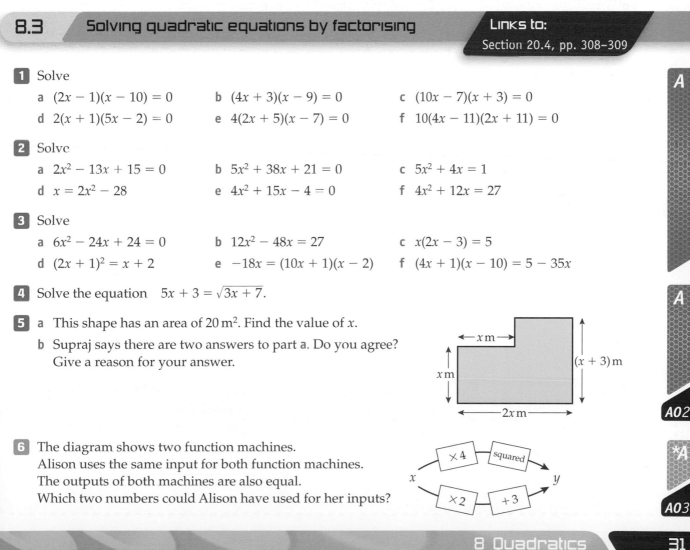

A02

6 The diagram shows two function machines.
Alison uses the same input for both function machines.
The outputs of both machines are also equal.
Which two numbers could Alison have used for her inputs?

*A

A03

Challenge yourself

You won't encounter questions like this in the exam, but the underlying maths is covered in your GCSE course. **Have a go!**

The diagram shows a graph with equation $y = ax^3 + bx^2 + cx + d$.
Work out the values of a, b, c and d.

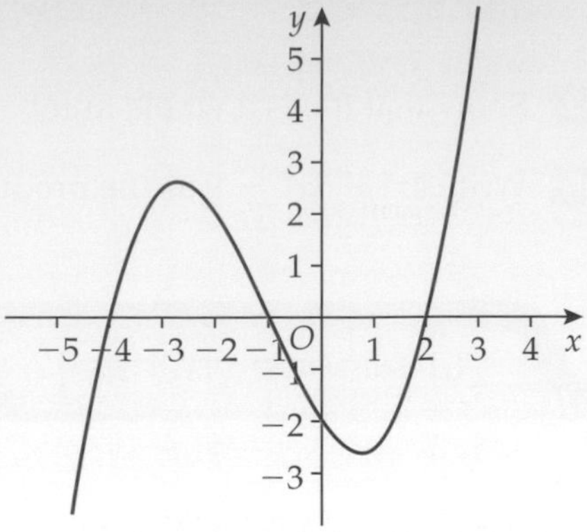

The points where the curve crosses the x-axis represent the solutions to the equation $y = 0$.

8.4 Using the quadratic formula

Links to:
Section 20.5, pp. 309–310

A

1 Use the quadratic formula to solve these equations. Leave your answers in surd form.

   a $x^2 - 6x - 5 = 0$ **b** $x^2 + 4x + 2 = 0$ **c** $x^2 - 12x + 15 = 0$

   d $-x^2 + 10x - 2 = 0$ **e** $2x^2 - 9x + 1 = 0$ **f** $5x^2 + 3x - 3 = 0$

2 Use the quadratic formula to solve these equations.

   a $8x^2 + 10x + 2 = 0$ **b** $\frac{1}{2}x^2 + 3x - 8 = 0$ **c** $2x^2 - \frac{1}{2}x - \frac{3}{2} = 0$

   d $3x^2 = 2x + 5$ **e** $-5x^2 = \frac{3}{2}x - 2$ **f** $11x^2 + 8x = 3$

3 Use the quadratic formula to solve these equations. Leave your answers in surd form.

   a $4x^2 = 1 - 5x$ **b** $3x^2 = 4x + 1$ **c** $x(x - 10) = 3$

   d $(x + 1)^2 = 3 - x$ **e** $(2x + 1)(x - 3) = 5$ **f** $(x + 2)(x + 6) = (2x + 3)^2$

A

4 Show that the solutions of the equation $2x^2 - 6x - 1 = 0$ are $x = \dfrac{3 \pm \sqrt{11}}{2}$

AO2

5 Explain why you cannot solve the equation $x^2 + 2x + 2 = 0$ using the quadratic formula.

A*

6 You are constructing an open-topped box from a square piece of paper.
You will cut square pieces off each corner to make a net.
The box must have a height of 2 cm.
If the volume of the finished box is 54 cm³, calculate the original width of the piece of paper, x cm. Give your answer in surd form.

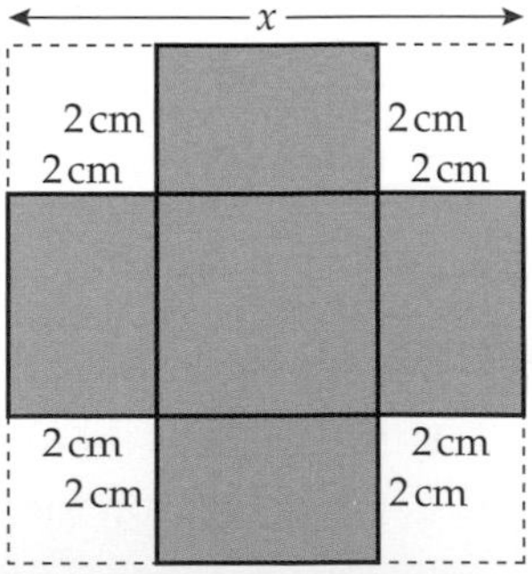

AO3

Challenge yourself

You won't encounter questions like this in the exam, but the underlying maths is covered in your GCSE course. **Have a go!**

The ratio of $a : b$ in this line is the same as the ratio $b : c$.

Show that this ratio is $\dfrac{1 + \sqrt{5}}{2} : 1$

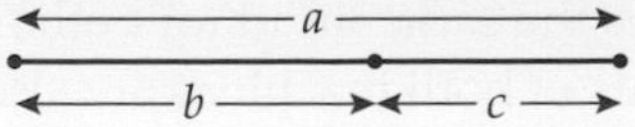

1 Use the discriminant to determine whether each quadratic equation has one solution, two solutions or no solutions.

a $x^2 - 2x - 1 = 0$ b $3x^2 + x + 1 = 0$ c $4x^2 - 8x + 4 = 0$

d $-x^2 + 2x - 5 = 0$ e $x^2 - 6x + 9 = 0$ f $-x^2 + x + 10 = 0$

*A

2 Write down the number of roots of each quadratic equation.

a $x^2 = 3x - 10$ b $10x = 25 + x^2$ c $4x = x(x - 3)$

d $(2x - 1)^2 + 6 = 0$ e $(x + 5)(x - 5) = 2$ f $(3x - 4)^2 = 6x - 9$

3 Julia says that if all the numbers in a quadratic equation are positive then the equation has no solutions. Is Julia correct? Give a reason for your answer.

*A

4 Use the discriminant to match each quadratic equation to its graph.

a $y = x^2 - 6x + 9$ b $y = 2x^2 - x + 3$ c $y = x^2 + 4x + 2$

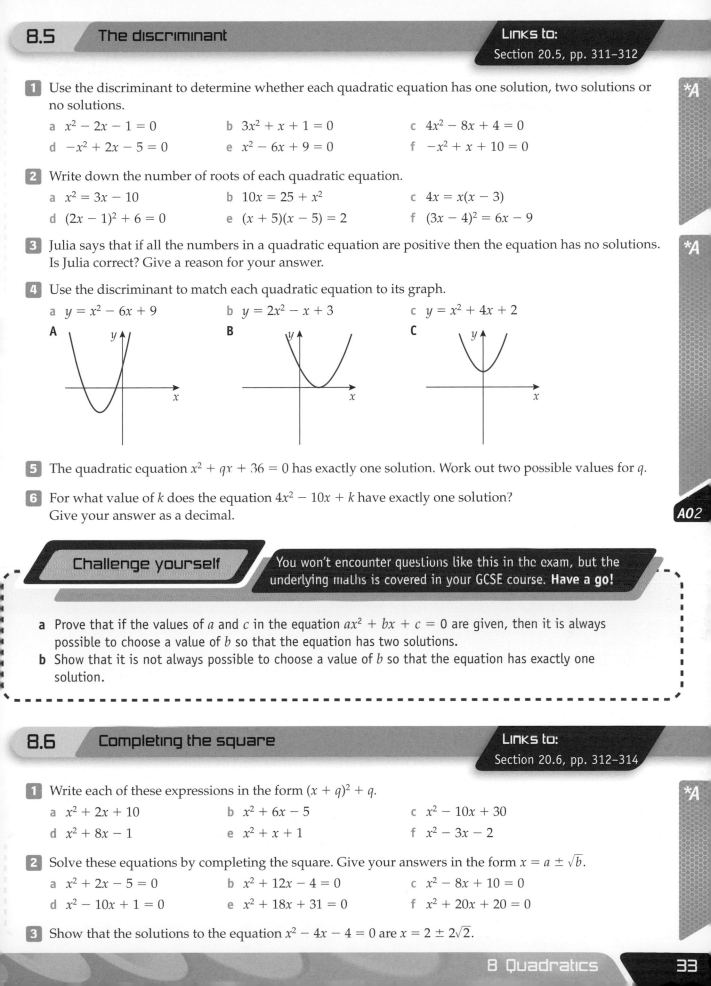

5 The quadratic equation $x^2 + qx + 36 = 0$ has exactly one solution. Work out two possible values for q.

6 For what value of k does the equation $4x^2 - 10x + k$ have exactly one solution?
Give your answer as a decimal.

AO2

Challenge yourself

You won't encounter questions like this in the exam, but the underlying maths is covered in your GCSE course. **Have a go!**

a Prove that if the values of a and c in the equation $ax^2 + bx + c = 0$ are given, then it is always possible to choose a value of b so that the equation has two solutions.

b Show that it is not always possible to choose a value of b so that the equation has exactly one solution.

1 Write each of these expressions in the form $(x + q)^2 + q$.

a $x^2 + 2x + 10$ b $x^2 + 6x - 5$ c $x^2 - 10x + 30$

d $x^2 + 8x - 1$ e $x^2 + x + 1$ f $x^2 - 3x - 2$

*A

2 Solve these equations by completing the square. Give your answers in the form $x = a \pm \sqrt{b}$.

a $x^2 + 2x - 5 = 0$ b $x^2 + 12x - 4 = 0$ c $x^2 - 8x + 10 = 0$

d $x^2 - 10x + 1 = 0$ e $x^2 + 18x + 31 = 0$ f $x^2 + 20x + 20 = 0$

3 Show that the solutions to the equation $x^2 - 4x - 4 = 0$ are $x = 2 \pm 2\sqrt{2}$.

A* **4** Solve the equation $4x^2 - 4x - 1 = 0$ by completing the square.

You will need to divide through by 4 first.

AO2

A* **5** A quadratic equation has roots given by $x = 5 \pm \sqrt{7}$.
Write this equation in the form $ax^2 + bx + c = 0$ where a, b and c are integers.

6 a Write $x^2 + 6x + 20$ in completed square form.

b Use your answer to part **a** to calculate the smallest possible value of $x^2 + 6x + 20$.

c For what value of x does this smallest value occur?

AO3 **7** If $4x^2 + 12x + a = (2x + b)^2$, find the values of a and b.

Challenge yourself

You won't encounter questions like this in the exam, but the underlying maths is covered in your GCSE course. **Have a go!**

a By completing the square, show that the solutions to the equation $x^2 + 2bx + c = 0$ are given by
$x = -b \pm \sqrt{b^2 - c}$

b Prove that the solutions to the equation $ax^2 + 2bx + c = 0$ are given by $x = -\dfrac{b}{a} \pm \sqrt{\dfrac{b^2 - ac}{a^2}}$

Further algebraic methods

Key Points

Simultaneous equations **A***

You can solve one quadratic equation and one linear equation simultaneously using substitution. The solutions correspond to the points where the graphs of the two equations cross.

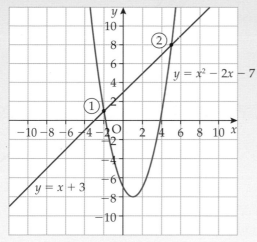

There are two solutions to the simultaneous equations
$$y = x^2 - 2x - 7$$
$$y = x + 3$$
① $\quad x = -2, y = 1$
② $\quad x = 5, y = 8$

Algebraic fractions **A** **A***

You can simplify algebraic fractions by factorising the numerator and the denominator.
$$\frac{x^2 - 9}{x^2 + 8x + 15} = \frac{(x + 3)(x - 3)}{(x + 3)(x + 5)}$$
$$= \frac{x - 3}{x + 5}$$

You can add and subtract algebraic fractions by writing them over a common denominator.
$$\frac{1}{4x} + \frac{2x}{6y} = \frac{3y}{12xy} + \frac{4x^2}{12xy}$$
$$= \frac{3y + 4x^2}{12xy}$$

> **The lowest common multiple of $4x$ and $6y$ is $12xy$.**

Solving equations with algebraic fractions **A** **A***

You can simplify equations with algebraic fractions by multiplying by the denominators.
$$\frac{1}{x} - \frac{2}{x + 1} = 1$$
$$1 - \frac{2x}{x + 1} = x$$
$$(x + 1) - 2x = x(x + 1)$$
$$x^2 + 2x - 1 = 0$$
$$x = -1 \pm \sqrt{2}$$

> **Multiply through by x then multiply through by $x + 1$.**

> **This is a quadratic equation. Solve it by completing the square.**

Changing the subject of a formula **A** **A***

If a letter appears twice in a formula:

- rearrange the formula so all the terms containing the letter are on the same side

- factorise to get the letter on its own.
$$y = \frac{10x}{x - 1}$$
$$y(x - 1) = 10x$$
$$xy - 10x = y$$
$$x(y - 10) = y$$
$$x = \frac{y}{y - 10}$$

> **Factorise to get x on its own.**

> **Divide both sides by $y - 10$.**

9.1 Simultaneous equations . . . one linear, one quadratic

Links to:

Section 21.1, pp. 317–320

1 Solve these simultaneous equations using substitution.

a $\quad y = x + 6$
$\quad\; y = x^2$

b $\quad y = 2x$
$\quad\; y = x^2 - 3$

c $\quad y = 10x$
$\quad\; y = 2x^2$

d $\quad y^2 = 6x$
$\quad\; y = 2x - 6$

> *A
> **Substitute the linear equation for y into the quadratic equation and solve to find all the possible values of x. Next, find the corresponding values of y.**

2 Solve these simultaneous equations.

a $y = x + 1$

$y = x^2 + 6x - 5$

b $y = 2x + 1$

$y = x^2 - 6x + 16$

c $y = 5x$

$y = 3x^2 - 2$

d $y = 3x + 2$

$y^2 = 2x^2 - 1$

3 Solve these simultaneous equations using substitution.

$$xy = 14$$
$$4x + 3y = 29$$

4 The diagram shows the curve $y = \dfrac{7}{x}$ and the line $y = 10 - x$.

Find the coordinates of the points A and B where these lines intersect. Give your answers in surd form.

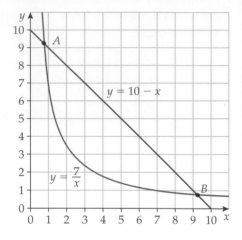

5 Two numbers m and n have a product of -18 and a sum of 10.5.

Write two equations and solve them simultaneously to find the values of m and n.

6 Solve these simultaneous equations. Give your answers in surd form.

$$a^2 + b = 6$$
$$2a + b = 4$$

7 a By considering the discriminant, show that these simultaneous equations have no solutions.

$$y + x = 8$$
$$y = 3 - 3x^2$$

b What does your answer to part a tell you about the graphs of these two equations?

Challenge yourself

You won't encounter questions like this in the exam, but the underlying maths is covered in your GCSE course. **Have a go!**

Find all the possible sets of values for a, b and c that satisfy these three equations simultaneously.

$$ab + c = 6$$
$$bc + a = 6$$
$$ac + b = 6$$

9.2 Simplifying algebraic fractions

Links to:

Section 21.2, pp. 321–324

1 Simplify each expression as much as possible.

a $\dfrac{x + 3}{x^2 + 5x + 6}$

b $\dfrac{2x - 10}{x^2 - 3x - 10}$

c $\dfrac{3x + 6}{x^2 + 4x + 4}$

d $\dfrac{x^2 + 9x + 8}{x^2 - 4x - 5}$

e $\dfrac{x^2 + 10x + 16}{x^2 + 3x - 40}$

f $\dfrac{2x^2 + 6x - 20}{x^2 - 6x + 8}$

2 Write each of these as a single fraction.

a $\dfrac{1}{ab} + \dfrac{1}{2b}$

b $\dfrac{2}{xy} - \dfrac{3}{x^2}$

c $\dfrac{2x}{4y} - \dfrac{y}{3x}$

3 a Factorise $4x^2 - 25$. **b** Simplify $\dfrac{4x^2 - 25}{2x^2 + 3x - 5}$

4 Simplify each expression as much as possible.

a $\dfrac{x^2 - 9}{x^2 + 8x + 15}$
b $\dfrac{2x^2 - 12x + 10}{x^2 - 25}$
c $\dfrac{2x^2 - 5x}{4x^2 - 4x - 15}$

5 Simplify each expression as much as possible.

a $\dfrac{x}{x^2 + 3x - 4} \div \dfrac{x^2 + 3x}{x - 1}$

b $\dfrac{x + 1}{x^2 + 7x + 12} \div \dfrac{x^2 + 2x + 1}{x + 4}$

c $\dfrac{9x^2 - 4}{x + 2} \div \dfrac{3x^2 + 5x + 2}{x^2 + x - 2}$

> **To divide by a fraction you turn it upside down and multiply.**
> $\frac{1}{3} \div \frac{3}{4} = \frac{1}{3} \times \frac{4}{3} = \frac{4}{9}$

6 Write each of these as a single fraction.

a $\dfrac{1}{x + 1} + \dfrac{2}{x - 1}$
b $\dfrac{5}{x - 3} - \dfrac{3}{x + 2}$
c $\dfrac{x}{2x - 1} + \dfrac{10}{x + 6}$

7 Write $2x + \dfrac{3}{x - 1}$ as a single fraction.

8 Prove that $\dfrac{1}{n + 1} + \dfrac{n + 1}{n - 1} = \dfrac{n(n + 3)}{n^2 - 1}$

Challenge yourself

You won't encounter questions like this in the exam, but the underlying maths is covered in your GCSE course. **Have a go!**

a and b are integers with $a \neq b$, $a \neq 0$ and $b \neq 0$.

Show that $\dfrac{a}{\dfrac{1}{b - a} - \dfrac{1}{b}}$ is also an integer.

9.3 Equations involving algebraic fractions

Links to:
Section 21.2, pp. 324–326

1 The first rectangle has been split in half.
The second rectangle has been split into fifths.
The two shaded sections have the same area.
Work out the value of x.

2 Angela has a budget of £x per week for a two-week holiday. In the first week of the holiday she spends £240 on car rental then divides the remainder equally between herself and her sister.
In the second week she spends £300 on her hotel bill then divides the remainder equally between herself, her sister and her niece.

a Write expressions for the amount of money Angela's sister receives each week.

b Angela's sister received one-third of the total holiday budget.
Write an equation and solve it to find the value of x.

3 Solve these equations.

a $\dfrac{6}{x(x+3)} + \dfrac{7}{2(x+3)} = \dfrac{8}{x}$

b $\dfrac{1}{(p+1)^2} + \dfrac{1}{2p(p+1)} = \dfrac{1}{5p(p+1)^2}$

4 The areas of these two rectangles are shown. Work out the value of h.

←$x+5$→←——$4x$——→

h | 20 cm² | 30 cm²

5 Solve

a $\dfrac{5}{x} + \dfrac{2}{x+2} = 3$

b $\dfrac{1}{x+3} + \dfrac{3}{3x+1} = 1$

c $\dfrac{6}{x-1} + \dfrac{8}{x} = 4$

d $\dfrac{4}{3x-1} + \dfrac{5}{x-1} = 3$

e $\dfrac{8}{x+3} - \dfrac{5}{x+2} = -1$

f $\dfrac{10}{2x-5} - \dfrac{5}{x-4} = -3$

6 Show that the equation $\dfrac{1}{x+5} - \dfrac{2}{x+3} = 1$ has no solutions.

7 Solve the equation $\dfrac{2}{x} + \dfrac{2}{x+1} = 1$. Give your answer in surd form.

8 One number is four times as big as another number. The sum of their reciprocals is $\dfrac{1}{4}$.
Work out the two numbers.

9 Solve

a $\dfrac{3}{x} - \dfrac{3}{x+3} = \dfrac{9}{10}$

b $\dfrac{1}{2x-1} + \dfrac{x+1}{3x} = \dfrac{5}{6}$

c $\dfrac{1+x}{x-2} - \dfrac{x-4}{x} = 1$

10 The area of the first rectangle is 5 cm².
The area of the second rectangle is 8 cm².
The distance AB is 1.75 cm. Find the value of x.

A ←$x-4$→

1.75 cm | Area = 5 cm²

Area = 8 cm²

B ←——$2x$——→

Challenge yourself

You won't encounter questions like this in the exam, but the underlying maths is covered in your GCSE course. **Have a go!**

Show that you can write $\dfrac{3x+1}{x^2-x-6}$ in the form $\dfrac{A}{x-3} + \dfrac{B}{x+2}$, where A and B are whole numbers.

9.4 Changing the subject of a formula

Links to:
Section 21.3, pp. 326–328

1 Make n the subject of these formulae.

a $an + 2 = b + cn$

b $2n + x^2 = mn - 1$

c $a(n+m) = b(n-m)$

2 Make c the subject of the formula $a\sqrt{c^2+1} = b\sqrt{c^2+5}$.

3 The formula for the area of a trapezium is $A = \dfrac{ah + bh}{2}$

   a Rearrange this formula to make h the subject.

   b Find the height of the trapezium below.

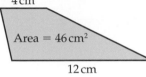

   Area = 46 cm²

   4 cm

   12 cm

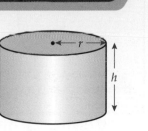

4 Rearrange these formulae to make P the subject.

   a $X = \dfrac{10P}{P + Q}$ **b** $Y = \dfrac{P + 1}{P - 2}$ **c** $Z = \dfrac{P + Q + R}{3PQR}$

5 Rearrange this formula to make x the subject.

   $\dfrac{px + 1}{b} = \dfrac{x}{a} + \dfrac{x + q}{c}$

6 Rearrange this formula to make y the subject.

   $\dfrac{a}{y^2 + 1} = \dfrac{b}{2y^2}$

7 Make R the subject of the formula $T = \sqrt{\dfrac{2\pi R}{R + h}}$

8 The ratio of two consecutive even numbers is given by the formula $R = \dfrac{n + 2}{n}$

   a Rearrange this formula to make n the subject.

   b Find the value of n if $R = 1.05$.

   > **Write R as a fraction first.**

Challenge yourself

You won't encounter questions like this in the exam, but the underlying maths is covered in your GCSE course. **Have a go!**

The formula for the surface area of a cylinder is $A = 2\pi r^2 + 2\pi rh$.

a Rearrange this formula to make r the subject.

b Explain why there is always exactly one possible radius for a given height and surface area.

Links to:
Higher Student Book
Ch26, pp. 385–386,
Ch27, pp. 398–405

Key Points

Pyramids [A]

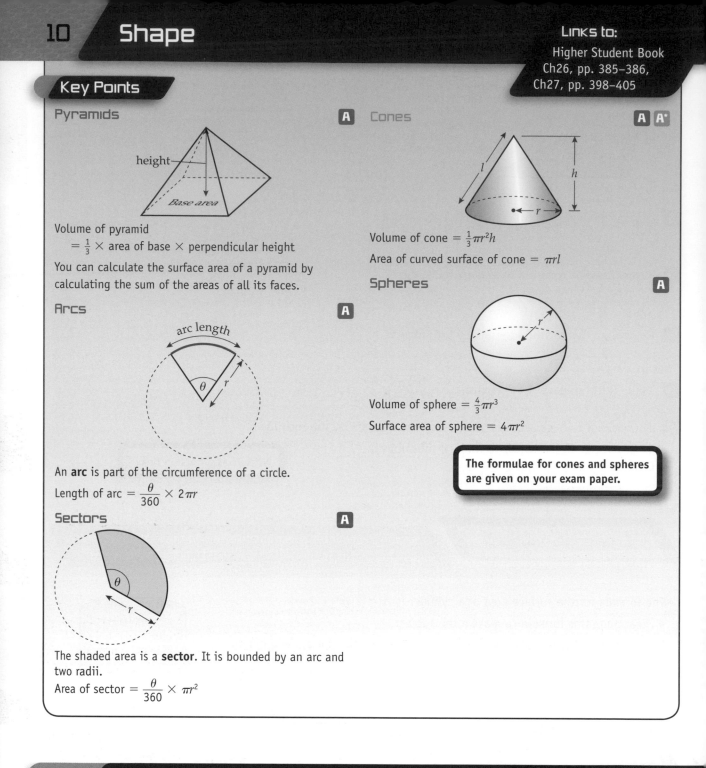

height—
Base area

Volume of pyramid
$= \frac{1}{3} \times$ area of base $\times$ perpendicular height

You can calculate the surface area of a pyramid by calculating the sum of the areas of all its faces.

Arcs [A]

arc length

θ r

An **arc** is part of the circumference of a circle.

Length of arc $= \frac{\theta}{360} \times 2\pi r$

Sectors [A]

θ

r

The shaded area is a **sector**. It is bounded by an arc and two radii.

Area of sector $= \frac{\theta}{360} \times \pi r^2$

Cones [A] [A*]

l h

r

Volume of cone $= \frac{1}{3}\pi r^2 h$

Area of curved surface of cone $= \pi r l$

Spheres [A]

r

Volume of sphere $= \frac{4}{3}\pi r^3$

Surface area of sphere $= 4\pi r^2$

The formulae for cones and spheres are given on your exam paper.

10.1   Pyramids

Links to:
Section 26.5, pp. 385–386

A

1 Calculate the volume and the total surface area of each pyramid.

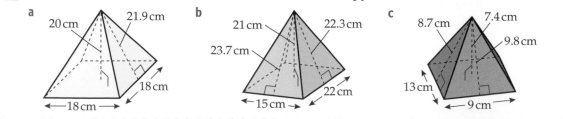

a
20 cm   21.9 cm

18 cm

←18 cm→

b
21 cm   22.3 cm

23.7 cm

←15 cm→   22 cm

c
8.7 cm   7.4 cm

9.8 cm

13 cm   ←9 cm→

2 The diagram shows a triangular based pyramid.

All the faces are equilateral triangles.

Calculate

  a the volume of the pyramid

  b the total surface area of the pyramid.

8.7 cm

8.2 cm

10 cm

A

> **This shape is called a regular tetrahedron.**

3 This diagram shows a square-based pyramid.
Vertex A is directly above vertex B.

Calculate the volume of this pyramid.

A

24 cm

C

D

B

17 cm

E

4 The great pyramid of Giza in Egypt is 146 m tall. Its base is a square with sides of length 230 m.

  a Calculate the volume of the pyramid in cubic metres. Give your answer to 3 significant figures.

  b The pyramid is made from stone with a density of 1840 kg/m^3.
    Calculate the approximate mass of the pyramid. Give your answer in standard form.

A

5 A child's building block is made from a cuboid and a pyramid.
The vertex of the pyramid is directly above the centre of the base.

Calculate

  a the volume of the block

  b the surface area of the block.

Give your answers correct to 2 decimal places.

3.3 cm   6.5 cm

5.6 cm

4 cm

10 cm

4.2 cm

6 A pyramid has a square base of side length 4 cm and a volume of 64 cm^3.
Calculate its vertical height.

AO2

7 The diagram shows a square-based pyramid with its top cut off.
Calculate its volume.

> **Use similar triangles to calculate the height of the original pyramid.**

*A

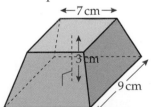

←7 cm→

3 cm

9 cm

AO3

Challenge yourself

You won't encounter questions like this in the exam, but the underlying maths is covered in your GCSE course. **Have a go!**

The diagram shows a square-based pyramid with side length 12 cm and height 20 cm. A vertical slice has been made parallel to one edge of the base of the pyramid at a distance 3 cm from the edge.

Calculate the volume of the new shape.

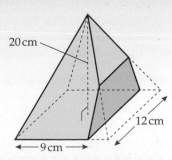

20 cm

12 cm

←9 cm→

Links to:
Section 27.3, pp. 398–400

A

1 Calculate the lengths marked x in these circles. Give your answers in terms of π where necessary.

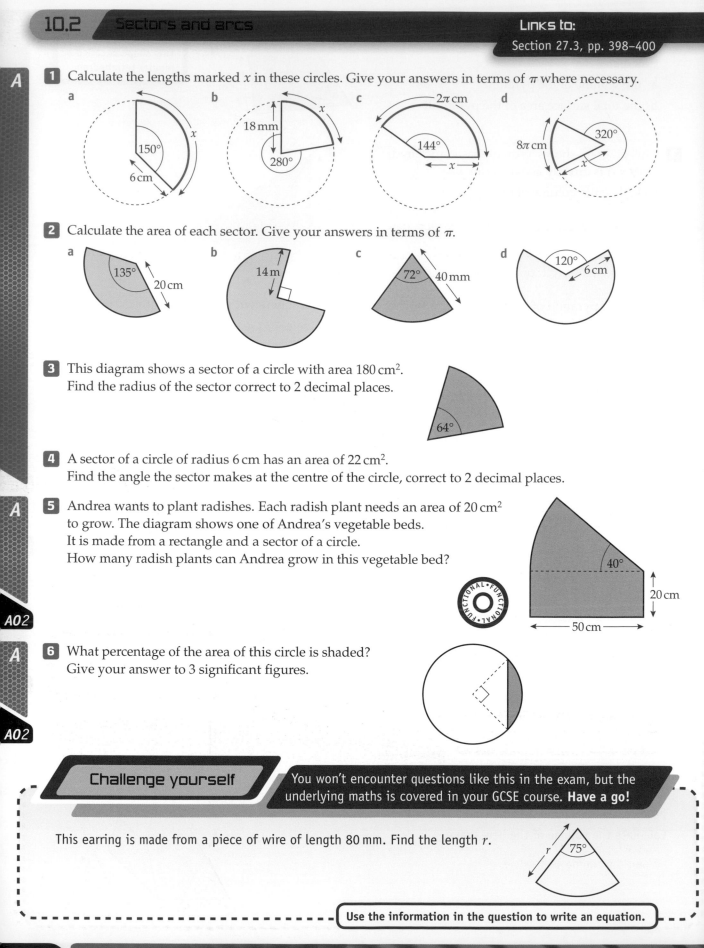

a

b

c 2π cm

d

2 Calculate the area of each sector. Give your answers in terms of π.

a   135°   20 cm

b   14 m

c   72°   40 mm

d   120°   6 cm

3 This diagram shows a sector of a circle with area 180 cm².
Find the radius of the sector correct to 2 decimal places.

64°

4 A sector of a circle of radius 6 cm has an area of 22 cm².
Find the angle the sector makes at the centre of the circle, correct to 2 decimal places.

A

5 Andrea wants to plant radishes. Each radish plant needs an area of 20 cm²
to grow. The diagram shows one of Andrea's vegetable beds.
It is made from a rectangle and a sector of a circle.
How many radish plants can Andrea grow in this vegetable bed?

40°

20 cm

50 cm

AO2

A

6 What percentage of the area of this circle is shaded?
Give your answer to 3 significant figures.

AO2

Challenge yourself

You won't encounter questions like this in the exam, but the
underlying maths is covered in your GCSE course. **Have a go!**

This earring is made from a piece of wire of length 80 mm. Find the length r.

r 75°

Use the information in the question to write an equation.

Links to:
Section 27.4, pp. 401–403

1   Calculate the volume and the surface area of each cone. Give your answers correct to 3 significant figures.

a   10 cm   10.8 cm   4 cm

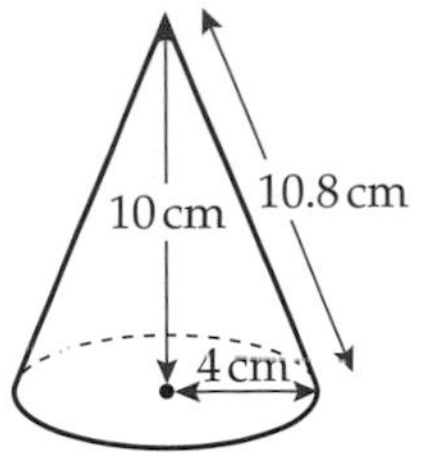

b   2 mm   9.2 mm   9 mm

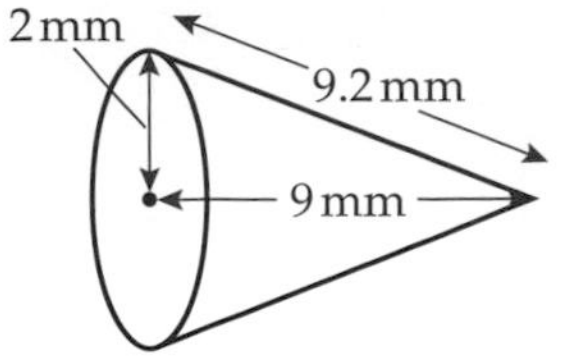

c   3.8 cm   3.2 cm   5 cm

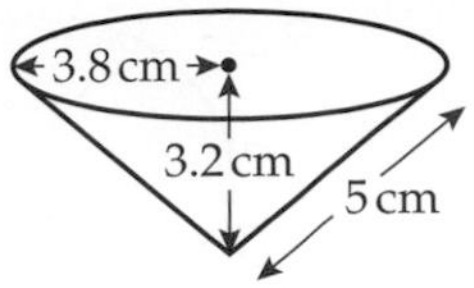

d   10 m   12 m   6.6 m

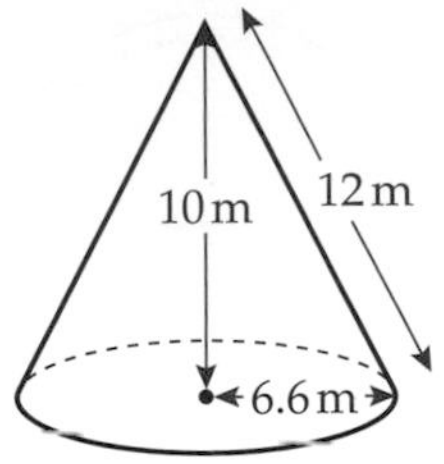

2   A cone has volume $375\pi\,\text{cm}^3$ and vertical height 20 cm. Calculate the radius of its base.

3   A metal cube of side length 8 cm is melted down. The metal is used to make a cone with radius 6 cm. Calculate the height of the cone correct to 3 significant figures.

4   The diagram shows a tent constructed from a triangular prism and half a cone. Calculate the total volume of the tent correct to 3 significant figures.

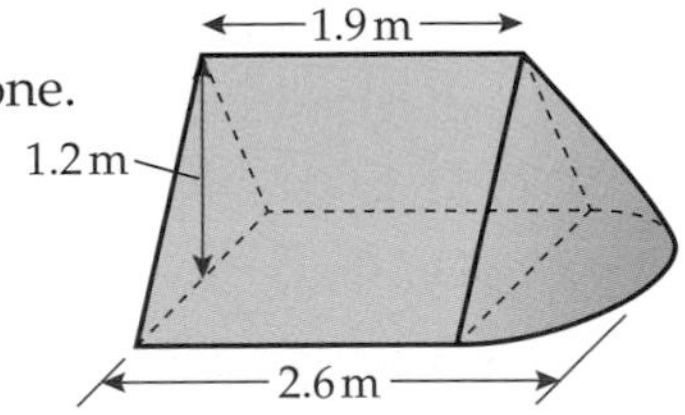

1.9 m   1.2 m   2.6 m

5   The diagram shows the net of a cone.

Calculate

a   the radius of the cone

b   the total surface area of the cone.

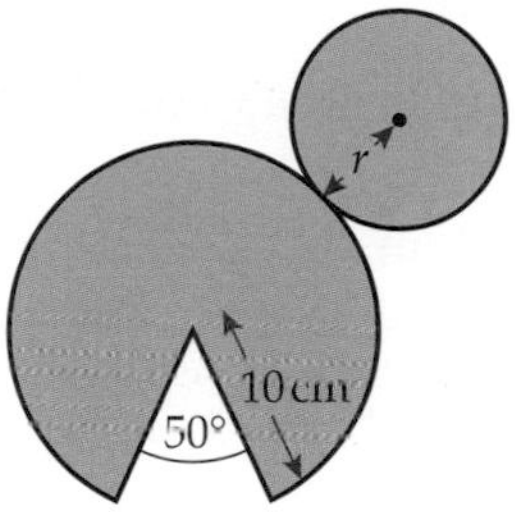

r 10 cm   50°

6   Karl has made a horizontal cut half way up this cone.
Find the ratio of the volume of the smaller piece to the volume of the larger piece.

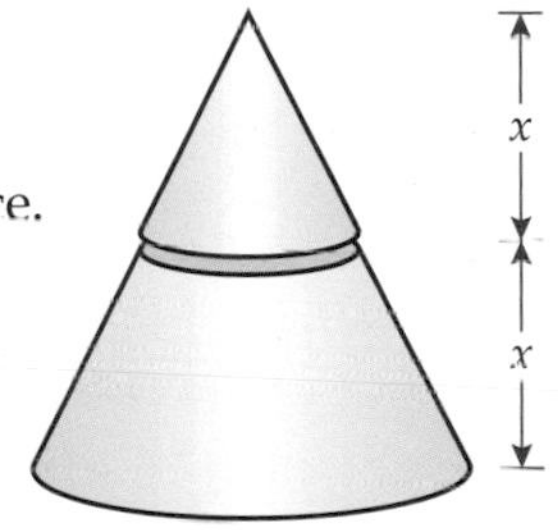

x x

7   A plinth for a statue is made by cutting the top off a cone.
Calculate the volume of the plinth.
Give your answer in terms of π.

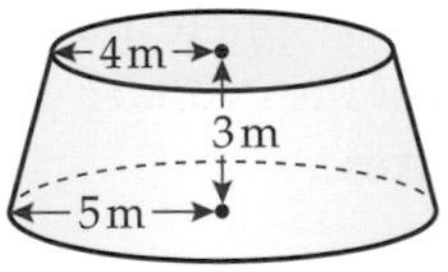

4 m   3 m   5 m

> This shape is called the frustum of a cone. Use similar triangles to work out the height of the original cone.

Challenge yourself

You won't encounter questions like this in the exam, but the underlying maths is covered in your GCSE course. **Have a go!**

The diagram shows an egg timer made from two identical glass cones of height 6 cm. When the egg timer is turned over the sand reaches 4 cm up the top cone.
Calculate the distance up the bottom cone that the sand reaches when the timing is finished.

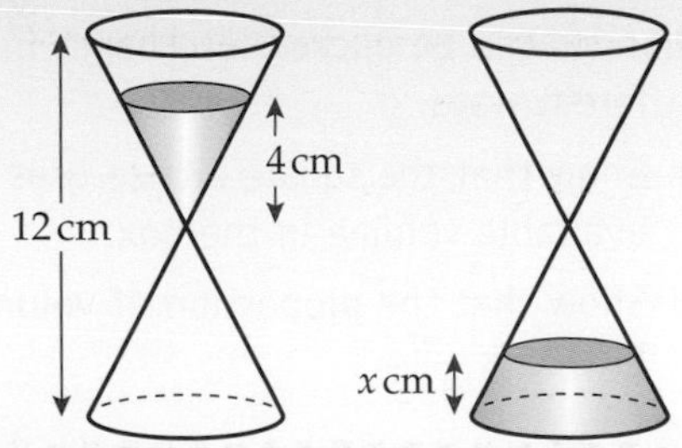

12 cm   4 cm x cm

A   A   AO2   A   AO3   *A   AO3

A **1** Calculate the volume and the surface area of each sphere . Give your answers in terms of π.

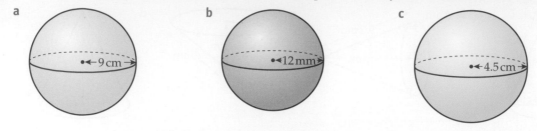

a   ●←9 cm→   b   ●←12 mm→   c   ●←4.5 cm→

2 A sphere has a volume of $80\,m^3$. Calculate its surface area, correct to 3 significant figures.

A **3** A Christmas pudding is in the shape of a sphere with volume $288\pi\,cm^3$. Work out the length of ribbon needed to wrap around the largest circumference of the pudding. Give your answer in terms of π.

4 The diagram shows a rolling pin constructed from a cylinder and two hemispheres.

Calculate

a   its volume

b   its surface area.

Give your answers correct to 3 significant figures.

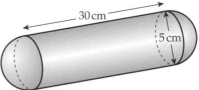

30 cm   5 cm

AO2

A **5** An ice cream cone has a radius of 3 cm and a height of 10 cm.
A ball of ice cream with radius 3 cm is placed in the cone and allowed to melt.

Show that the ice cream cone will overflow.

3 cm   10 cm

6 A cylindrical tube is constructed so that a sphere and a cone both fit exactly inside. Show that the volume of the cylinder is equal to the sum of the volumes of the sphere and the cone.

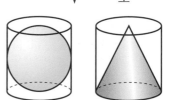

AO3

Oranges can be packed into boxes in a single layer in two different ways.

a Show that the square lattice uses approximately 52% of the available volume in the box.

b Show that the proportion of volume used in the hexagonal lattice is $\dfrac{\pi}{3\sqrt{3}}$

Square lattice   Hexagonal lattice

Key Points

Accuracy in calculations   **A**

To find the **greatest possible value** of a calculation you need to use:

- the upper bound of any values you add or multiply by
- the lower bound of any values you subtract or divide by.

To find the **least possible value** of a calculation you need to use

- the lower bound of any values you add or multiply by
- the upper bound of any values you subtract or divide by.

Solving problems involving accuracy   **A** **A***

When values have been rounded you sometimes need to choose the upper or the lower bound for a calculation.

Calculations involving density and speed   **A**

You can use the formulae for density and speed to solve problems involving upper and lower bounds.

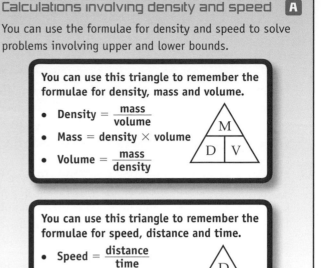

You can use this triangle to remember the formulae for density, mass and volume.

- Density = $\dfrac{\text{mass}}{\text{volume}}$
- Mass = density $\times$ volume
- Volume = $\dfrac{\text{mass}}{\text{density}}$

You can use this triangle to remember the formulae for speed, distance and time.

- Speed = $\dfrac{\text{distance}}{\text{time}}$
- Distance = speed $\times$ time
- Time = $\dfrac{\text{distance}}{\text{speed}}$

11.1   Density and speed   **Links to:**
Sections 28.2, 28.3, pp. 413–417

1 A snowboarder travels 120 m in 15 seconds. Both measurements are correct to 2 significant figures. Calculate the greatest possible value for her speed.   **A**

2 Polystyrene foam has a density of $28\,\text{kg/m}^3$ (to the nearest whole number).
An online shop orders 300 kg of foam, correct to 2 significant figures.
Calculate the least possible volume for the order.

3 The diagram shows a spherical steel ball-bearing.   **A**
The radius has been given correct to 1 decimal place.
The ball-bearing has a mass of 54 grams (to the nearest gram).

1.2 cm

Calculate the greatest possible value for the density of the steel used to make the ball-bearing.

Volume of sphere $= \frac{4}{3}\pi r^3$

4 The average speed of a skydiver is 50 m/s to the nearest ten metres per second.
He falls from a height of 3700 m (correct to 2 significant figures).
He must open his parachute at least 500 m from the ground.
He says that he can wait 60 seconds before opening his parachute.

   a Show that the skydiver is incorrect.

   b What is the longest he can safely wait before opening his parachute?
      Give your answer to 2 decimal places.

A02

5 The diagram shows a running track made up of two 100 m straight sections and two semicircular sections with diameter 64 m.

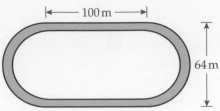

All measurements are correct to the nearest metre.
Sibtain runs around the track at an average speed of 7.2 m/s, correct to 1 decimal place.
Calculate an upper bound for the time taken by Sibtain to complete 4 laps.
Give your answer in minutes and seconds to the nearest second.

6 A rainwater tank contains 0.3 tonnes of water, correct to 1 decimal place.
The density of water is 1000 kg/m³, correct to 2 significant figures.
Charlie is using the water tank to fill up a bucket with a capacity of 8 litres, correct to the nearest litre.
He says he can fill the bucket at least 30 times.

a Show that Charlie is incorrect.

b How many times will Charlie definitely be able to fill the bucket?

Challenge yourself

You won't encounter questions like this in the exam, but the underlying maths is covered in your GCSE course. **Have a go!**

The density of iron is 7.9 g/cm³. The diagram shows a flask containing 350 ml of oil.

The capacity of the flask is 900 ml.
An iron bar of mass 4.2 kg is placed into the flask.
All measurements are given correct to 2 significant figures.
Show that there is a possibility that the flask will overflow.

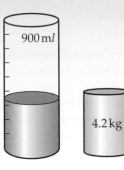

Links to:
Higher Student Book
Ch31, pp. 468–472,
Ch32, pp. 485–493

Key Points

Areas of similar objects

If one shape is an enlargement of another, the two shapes are **similar**.

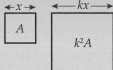

When a shape is enlarged by a linear scale factor k the area is increased by a factor k^2.

Enlarged area $= k^2 \times$ original area

Volumes of similar objects   **A**

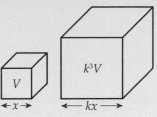

When a shape is enlarged by a linear scale factor k the volume is increased by a factor k^3.

Enlarged volume $= k^3 \times$ original volume

Congruent triangles   **A** **A***

Congruent shapes are exactly the same size and shape. Two triangles are congruent if one of the following conditions is satisfied.

- SSS: Three sides are equal

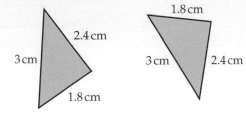

A
- SAS: Two sides and the included angle are equal

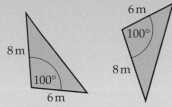

- ASA: Two angles and the included side are equal

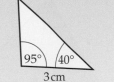
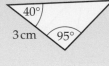

- AAS: Two angles and an opposite side are equal

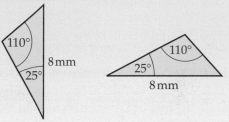

- RHS: A right angle, the hypotenuse and another side are equal

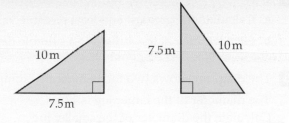

Negative scale factors   **A**

An enlargement with a negative scale factor produces an image that is on the other side of the centre of enlargement. The image appears upside down.

12.1 **Areas and volumes of similar objects**

Links to:
Section 32.3, pp. 485–490

1 The diagram shows two similar rectangles. The area of the smaller rectangle is $45\,cm^2$.
Calculate the area of the larger rectangle.

A

15 cm   20 cm

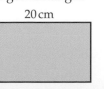

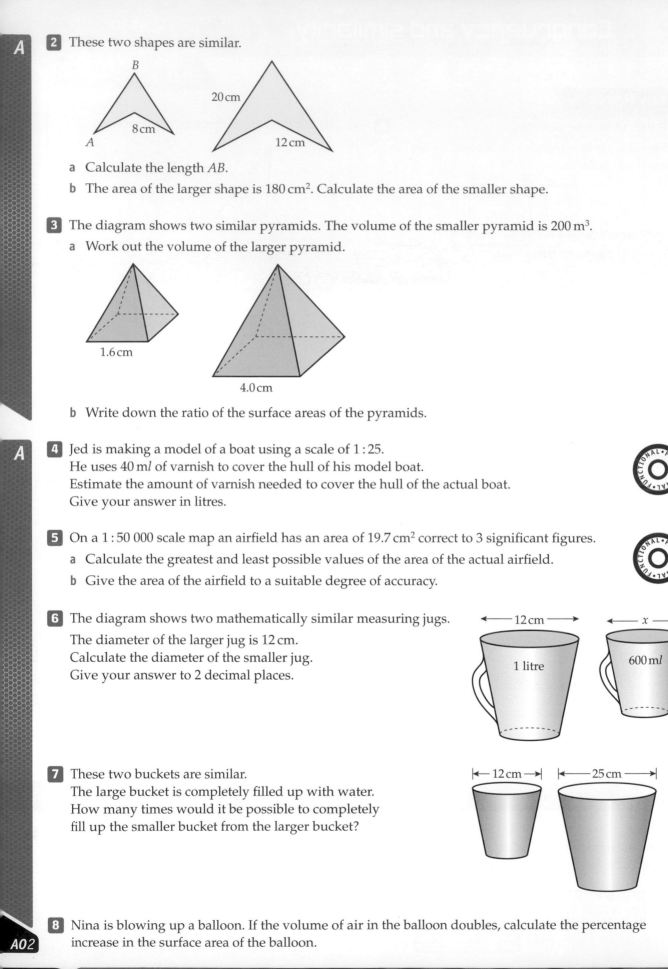

A

2 These two shapes are similar.

B

20 cm

8 cm

A

12 cm

a Calculate the length *AB*.

b The area of the larger shape is 180 cm². Calculate the area of the smaller shape.

3 The diagram shows two similar pyramids. The volume of the smaller pyramid is 200 m³.

a Work out the volume of the larger pyramid.

1.6 cm

4.0 cm

b Write down the ratio of the surface areas of the pyramids.

A

4 Jed is making a model of a boat using a scale of 1 : 25.
He uses 40 m*l* of varnish to cover the hull of his model boat.
Estimate the amount of varnish needed to cover the hull of the actual boat.
Give your answer in litres.

5 On a 1 : 50 000 scale map an airfield has an area of 19.7 cm² correct to 3 significant figures.

a Calculate the greatest and least possible values of the area of the actual airfield.

b Give the area of the airfield to a suitable degree of accuracy.

6 The diagram shows two mathematically similar measuring jugs.

The diameter of the larger jug is 12 cm.
Calculate the diameter of the smaller jug.
Give your answer to 2 decimal places.

← 12 cm → ← *x* →

1 litre

600 m*l*

7 These two buckets are similar.
The large bucket is completely filled up with water.
How many times would it be possible to completely
fill up the smaller bucket from the larger bucket?

← 12 cm → ← 25 cm →

8 Nina is blowing up a balloon. If the volume of air in the balloon doubles, calculate the percentage
increase in the surface area of the balloon.

A02

9 The diagram shows three similar dolls in different sizes.
The middle-sized doll has a volume of $162\,cm^3$ and a surface area of $180\,cm^2$.
Calculate

a the volume of the smallest doll

b the surface area of the largest doll.

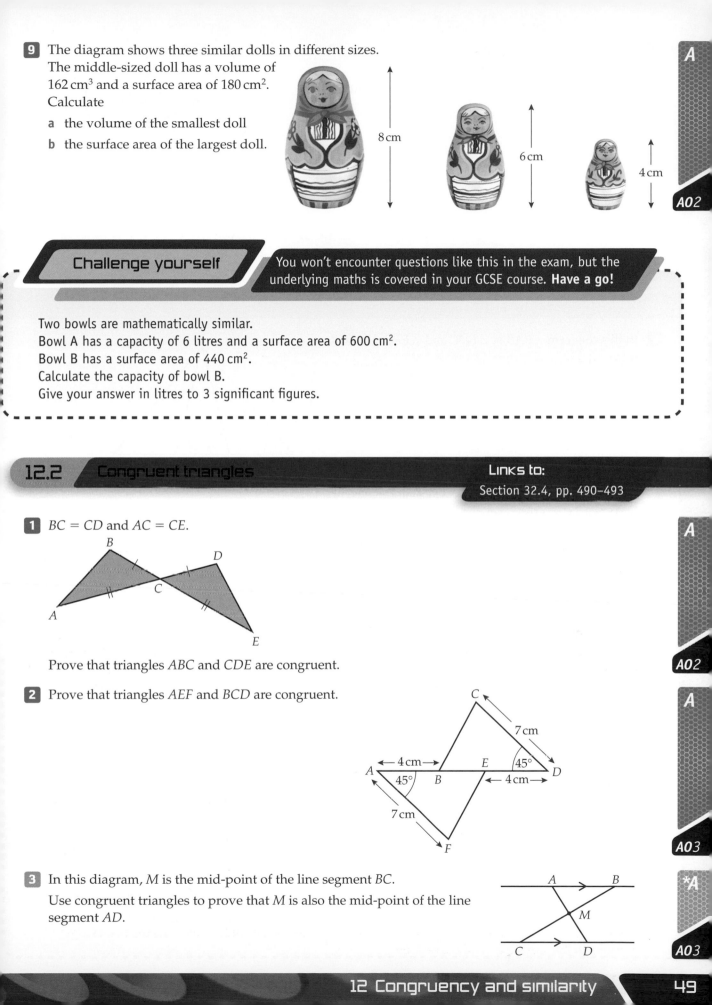

8 cm

6 cm

4 cm

Challenge yourself

You won't encounter questions like this in the exam, but the underlying maths is covered in your GCSE course. **Have a go!**

Two bowls are mathematically similar.
Bowl A has a capacity of 6 litres and a surface area of $600\,cm^2$.
Bowl B has a surface area of $440\,cm^2$.
Calculate the capacity of bowl B.
Give your answer in litres to 3 significant figures.

12.2   Congruent triangles

Links to:
Section 32.4, pp. 490–493

1 $BC = CD$ and $AC = CE$.

Prove that triangles ABC and CDE are congruent.

2 Prove that triangles AEF and BCD are congruent.

C

7 cm

←4 cm→   E   45°

A   45°   B   ←4 cm→   D

7 cm

F

3 In this diagram, M is the mid-point of the line segment BC.

Use congruent triangles to prove that M is also the mid-point of the line segment AD.

A   B

M

C   D

4 The diagram shows two circles with centres O and P.
The circles intersect at points X and Y.
Prove that triangles OXP and OYP are congruent.

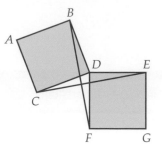

5 In this diagram, $ABCD$ and $DEFG$ are squares.
Prove that $BF = CE$.

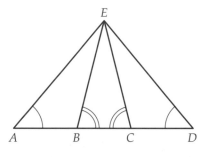

6 In this diagram $\angle EAB = \angle EDC$ and $\angle EBC = \angle ECB$.
Prove that triangles EBD and ECA are congruent.

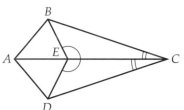

7 In this quadrilateral, CE bisects $\angle BCD$ and $\angle CEB = \angle CED$.
Prove that $BA = AD$.

Challenge yourself

You won't encounter questions like this in the exam, but the underlying maths is covered in your GCSE course. **Have a go!**

You can use congruent triangles to prove Pythagoras' theorem. This diagram shows a right-angled triangle XYZ. The squares $ABYX$, $CDZY$ and $EFZX$ have been drawn on the sides of the triangle.

a Prove that triangles AXZ and XYE are congruent.
b Explain why the area of triangle XYE is equal to half the area of rectangle $NEXM$.
c Prove that the area of square $ABYX$ is equal to the area of rectangle $NEXM$.
d Prove Pythagoras' theorem using this diagram.

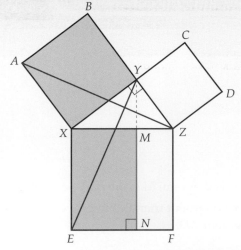

12 Congruency and similarity

1   Shape B is an enlargement of shape A.

    a   What is the scale factor of the enlargement?

    b   Work out the coordinates of the centre of enlargement.

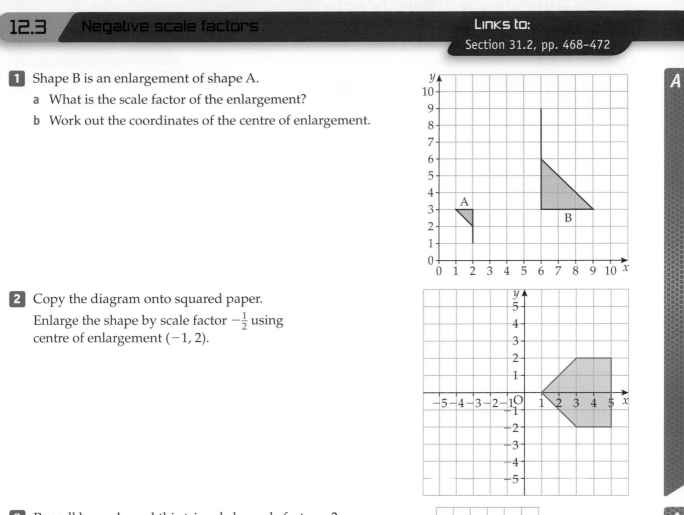

2   Copy the diagram onto squared paper.

Enlarge the shape by scale factor $-\frac{1}{2}$ using centre of enlargement $(-1, 2)$.

3   Russell has enlarged this triangle by scale factor -2 using C as the centre of enlargement.

Dom says that the inverse transformation is an enlargement by scale factor -2 using C as the centre of enlargement.

Is Dom correct? Show all of your working.

A

A02

Challenge yourself

You won't encounter questions like this in the exam, but the underlying maths is covered in your GCSE course. **Have a go!**

Fully describe the transformation which takes shape A onto shape B.

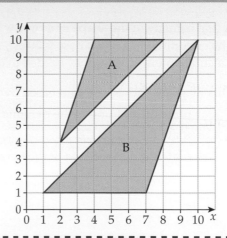

Key Points

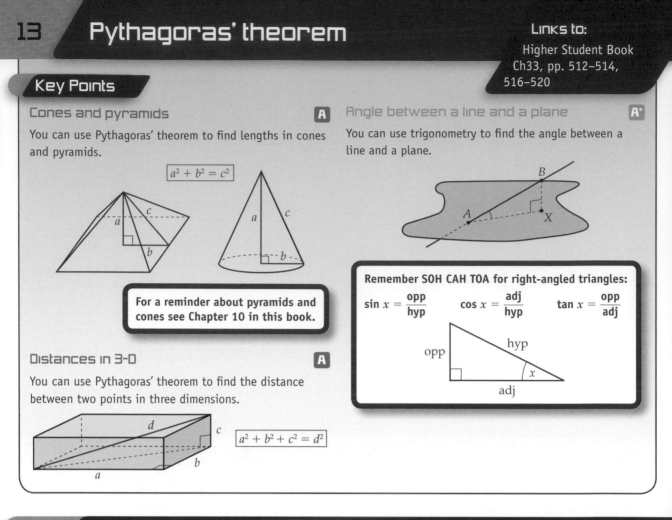

Cones and pyramids   [A]

You can use Pythagoras' theorem to find lengths in cones and pyramids.

$$a^2 + b^2 = c^2$$

> For a reminder about pyramids and cones see Chapter 10 in this book.

Distances in 3-D   [A]

You can use Pythagoras' theorem to find the distance between two points in three dimensions.

$$a^2 + b^2 + c^2 = d^2$$

Angle between a line and a plane   [A*]

You can use trigonometry to find the angle between a line and a plane.

> Remember SOH CAH TOA for right-angled triangles:
>
> $\sin x = \dfrac{\text{opp}}{\text{hyp}}$ $\cos x = \dfrac{\text{adj}}{\text{hyp}}$ $\tan x = \dfrac{\text{opp}}{\text{adj}}$

13.1   Pythagoras' theorem with pyramids and cones

Links to:
Section 33.4, pp. 512–514

Unless otherwise stated, give your answers correct to 3 significant figures.

[A]

1 Calculate the volume of this cone. Give your answer in terms of π.

25 cm

15 cm

2 A tent is made in the shape of a cone with height 2.2 m.
The circumference of the base is 10 m.
Calculate the area of canvas needed to construct the curved surface of the tent.

3 This diagram shows a square-based pyramid. Vertex E is directly above vertex D.

  a Calculate the volume of the pyramid.

  b Calculate the length CE.

  c Calculate the surface area of the pyramid.

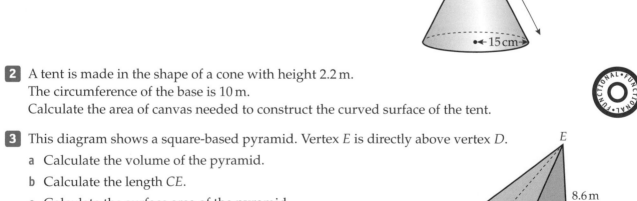

8.6 m

5.4 m

4 A cone has volume $550\,\text{cm}^3$. The height of the cone is $15\,\text{cm}$.
Work out

   a the radius of the cone

   b the slant height of the cone

   c the surface area of the cone.

A

5 Andrea is making a mobile.
She begins with a wire frame in the shape of a regular hexagon with side length $8\,\text{cm}$.
She attaches six pieces of string of length $17\,\text{cm}$ to the vertices of the hexagon and
ties them in a knot at the top.

   a Calculate the vertical height of her mobile.

   b Calculate the area of tissue paper she needs to cover the
sloping faces of her mobile.

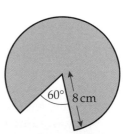

A02

6 Nisha cuts this sector out of cardboard and folds it up to make a cone-shaped cup.
What is the capacity of Nisha's cup?

> For help with areas of sectors
> see Chapter 10 in this book.

A03

Challenge yourself

You won't encounter questions like this in the exam, but the
underlying maths is covered in your GCSE course. **Have a go!**

An octahedron is a solid shape with eight faces. The diagram shows a regular
octahedron. All the faces are equilateral triangles with sides of length $2\,\text{cm}$.

Calculate the surface area and volume of this octahedron.
Give your answers in surd form.

2 cm

13.2   **Distances in 3-D**

Links to:

Section 33.6, pp. 516–518

1 The diagram shows a cuboid. $AB = 8\,\text{cm}$, $AC = 10\,\text{cm}$ and $BH = 12.5\,\text{cm}$.
Calculate

   a the distance AH

   b the distance AF.

Give your answers correct to 2 decimal places.

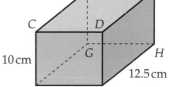
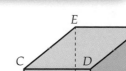
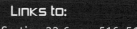

A

2 Calculate the length QV in this cuboid.

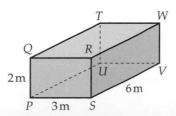

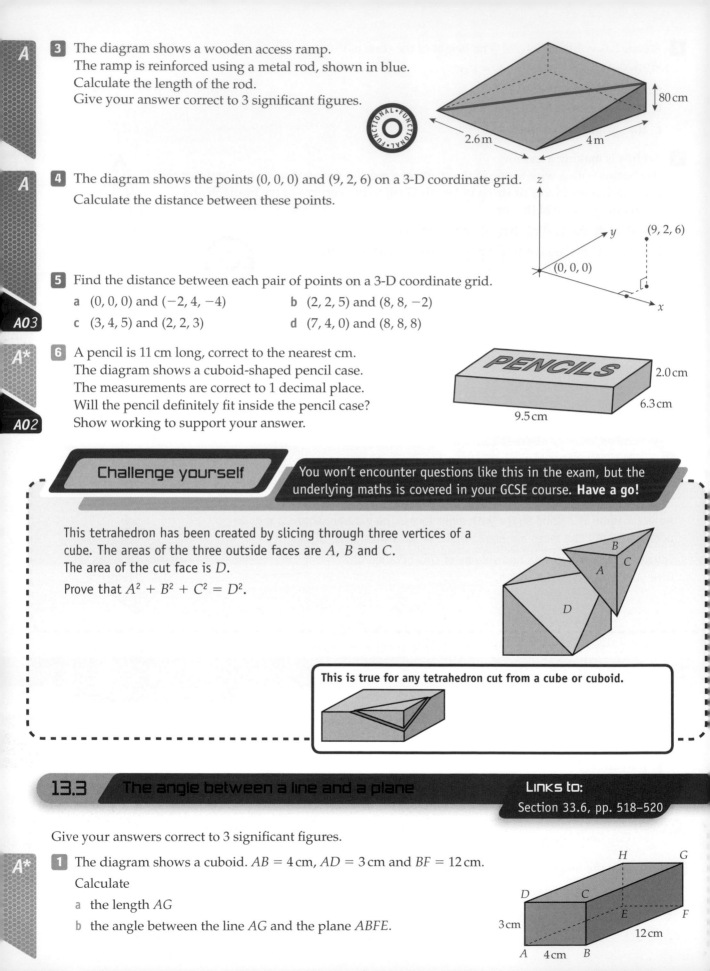

3 The diagram shows a wooden access ramp.
The ramp is reinforced using a metal rod, shown in blue.
Calculate the length of the rod.
Give your answer correct to 3 significant figures.

80 cm

2.6 m   4 m

4 The diagram shows the points (0, 0, 0) and (9, 2, 6) on a 3-D coordinate grid.
Calculate the distance between these points.

z

y   (9, 2, 6)

(0, 0, 0)

x

5 Find the distance between each pair of points on a 3-D coordinate grid.
  a (0, 0, 0) and (−2, 4, −4)   b (2, 2, 5) and (8, 8, −2)
  c (3, 4, 5) and (2, 2, 3)   d (7, 4, 0) and (8, 8, 8)

6 A pencil is 11 cm long, correct to the nearest cm.
The diagram shows a cuboid-shaped pencil case.
The measurements are correct to 1 decimal place.
Will the pencil definitely fit inside the pencil case?
Show working to support your answer.

PENCILS   2.0 cm
           6.3 cm
9.5 cm

Challenge yourself

You won't encounter questions like this in the exam, but the underlying maths is covered in your GCSE course. **Have a go!**

This tetrahedron has been created by slicing through three vertices of a cube. The areas of the three outside faces are A, B and C.
The area of the cut face is D.
Prove that $A^2 + B^2 + C^2 = D^2$.

B
C
A
D

This is true for any tetrahedron cut from a cube or cuboid.

13.3 The angle between a line and a plane

Links to:
Section 33.6, pp. 518–520

Give your answers correct to 3 significant figures.

1 The diagram shows a cuboid. $AB = 4$ cm, $AD = 3$ cm and $BF = 12$ cm.
Calculate
  a the length AG
  b the angle between the line AG and the plane $ABFE$.

H   G
D   C
      E   F
3 cm   12 cm
A 4 cm B

2 The diagram shows a triangular prism.
$AB = 7\,m$, $BC = 6\,m$ and $CF = 15\,m$.

Calculate

a the length AF

b the angle between the line AF and the plane $BCFE$.

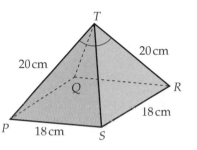

3 The diagram shows a square-based pyramid.
$AE = 19\,cm$ and $AD = 14\,cm$.
Vertex E is directly above vertex A.

Calculate the angle the line CE makes with the base of the pyramid.

4 The diagram shows a square-based pyramid $PQRST$.
The point T is directly above the centre of the square,
so that $PT = QT = RT = ST = 20\,cm$.

Calculate the size of the angle PTR.

5 A straight line is drawn between the points $(0, 0, 0)$ and $(4, 6, 3)$ on a three-dimensional coordinate grid.
Calculate the angle this line makes with the horizontal.

6 A river has two parallel banks. Nick is flying a kite from a point A on the north bank.
The length of the kite string is 60 m and the kite is directly above a point C on the
opposite bank of the river, 25 m downstream.
The kite string makes an angle of $22°$ with the ground.

Calculate

a the height of the kite

b the width of the river, AB.

Challenge yourself   You won't encounter questions like this in the exam, but the underlying maths is covered in your GCSE course. **Have a go!**

A set of integers (a, b, c) that satisfy the relationship $a^2 + b^2 = c^2$ is called a **Pythagorean triple.**

a Prove that if (a, b, c) is a Pythagorean triple then (ka, kb, kc) is also a Pythagorean triple.

b Prove that for any integer n, $(2n, n^2 - 1, n^2 + 1)$ is always a Pythagorean triple.

c Use the result in part **b** to find a right-angled triangle with integer sides whose hypotenuse has length 101 cm.

Key Points

Circle theorems

A **A***

You need to be able to use and prove these circle theorems.

Theorem 1: The angle subtended by an arc at the centre of a circle is twice the angle that it subtends at the circumference.

Theorem 2: Angles subtended by the same arc are equal.

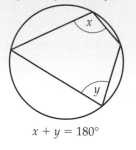

Theorem 3: Opposite angles in a cyclic quadrilateral are supplementary (they add up to 180°).

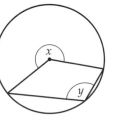

$$x + y = 180°$$

Theorem 4: The exterior angle of a cyclic quadrilateral is equal to the opposite interior angle.

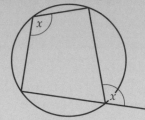

Theorem 5: The angle between a tangent and a chord is equal to the angle in the alternate segment.

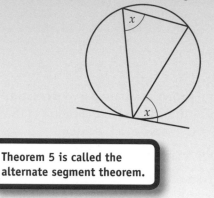

> **Theorem 5 is called the alternate segment theorem.**

14.1 Proofs of circle theorems

Links to:
Section 34.2, pp. 527–532

A

1 Prove that $x = 2y$.

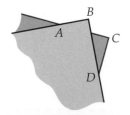

2 The diagram shows two overlapping pieces of paper.

   a Explain why the points A, B, C and D all lie on the circumference of the same circle.

   b What can you say about the line AD?

AO3

3 The diagram shows the path of an ice-skater on a circular ice rink.
The centre of the ice-rink is marked O.

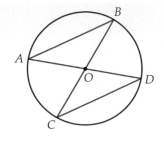

a Prove that $\angle ABC + \angle ADC = \angle AOC$

b Explain why all the angles at the edge of the circle are equal.

4 The diagram shows a circular pegboard.
XB is a diameter of the circle.
Rubber bands are attached to pegs X and Y
and stretched over other points on the circle.

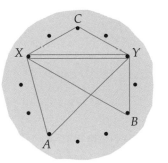

a Prove that $\angle XAY = \angle XBY$

b Prove that $\angle XAY + \angle XCY = 180°$

c Prove that $\angle XYB = 90°$

5 The diagram shows a cyclic quadrilateral.
Prove that $a + b = 180°$.

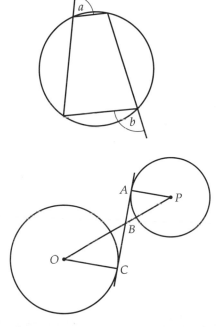

6 The line AC is a tangent to both circles.
Prove that triangles APB and OBC are similar.

Challenge yourself

You won't encounter questions like this in the exam, but the underlying maths is covered in your GCSE course. **Have a go!**

a Prove that triangles PBC and PDA are similar.
b Prove that $PA \times PB = PC \times PD$

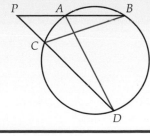

PA means the length of the line segment between P and A.

Links to:
Section 34.3, pp. 532–535

A

1 Calculate the size of each angle marked with a letter.

a

b

c

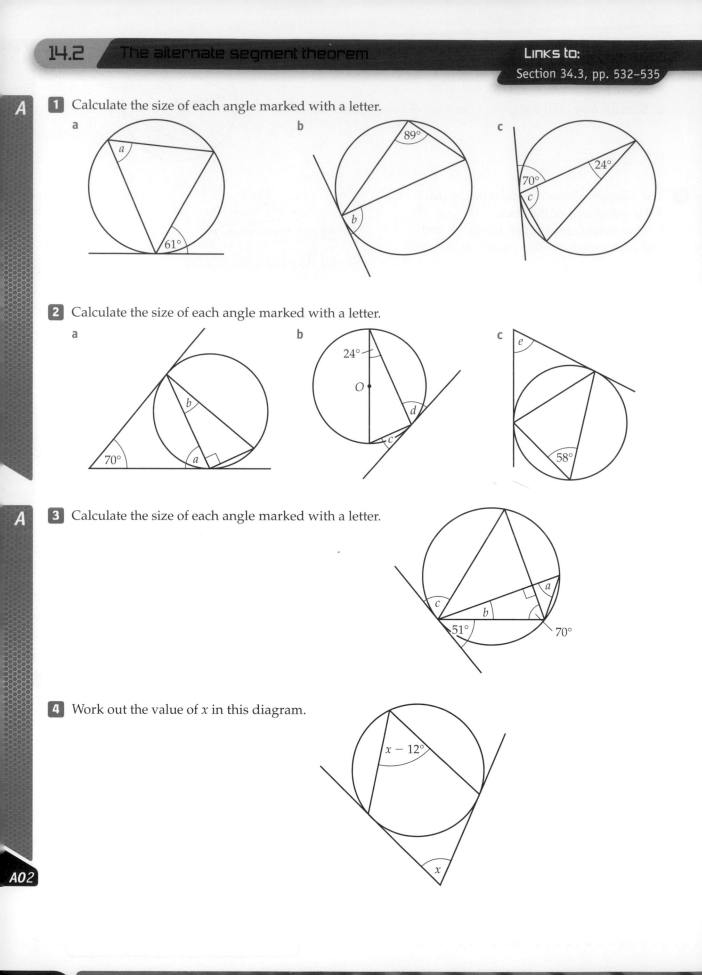

2 Calculate the size of each angle marked with a letter.

a

b

c

A

3 Calculate the size of each angle marked with a letter.

4 Work out the value of x in this diagram.

AO2

5 Calculate the size of each angle marked with a letter.

a
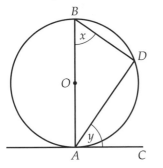

b
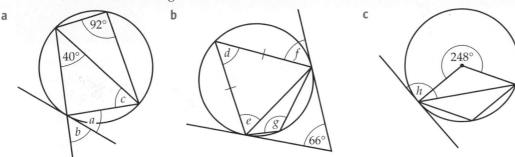

c

6 The diagram shows a tangent to a circle.

a Without using the alternate segment theorem, prove that $x = y$.

b Use your answer to part a to prove the alternate segment theorem.

Challenge yourself

You won't encounter questions like this in the exam, but the underlying maths is covered in your GCSE course. **Have a go!**

The points A, B, C and D all lie on the circumference of a circle with centre O.

Work out the size of angle CED.

Key Points

Solving quadratic equations graphically   A   A*

The solutions to the equation $2x^2 - 2x - 4 = 2 - x$ are the x-coordinates of the points where the graphs of $y = 2x^2 - 2x - 4$ and $y = 2 - x$ intersect.

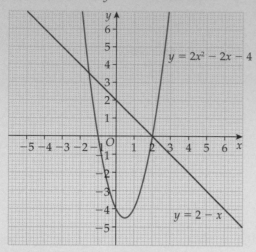

The solutions are $x = -1.5$ and $x = 2$.

Solving problems using quadratic equations   A*

Some problems give rise to quadratic equations.
If you can't factorise a quadratic equation you can use the quadratic formula. The solutions to the quadratic equation $ax^2 + bx + c = 0$ are given by the formula

$$x = \frac{-b \pm \sqrt{b^2 - 4ac}}{2a}$$

The expression $b^2 - 4ac$ is called the discriminant.

- If $b^2 - 4ac > 0$ there are two solutions.
- If $b^2 - 4ac = 0$ there is one solution.
- If $b^2 - 4ac < 0$ there are no solutions.

Graphs of reciprocal functions and combined functions   A   A*

A reciprocal function has the form $y = \frac{k}{x}$, where k is a constant. You can't work out the value of y if $x = 0$, so a reciprocal graph is not defined for $x = 0$.

If a function contains a linear or quadratic term together with a reciprocal term it is called a combined function. You can draw the graph of a combined function by drawing a table of values.

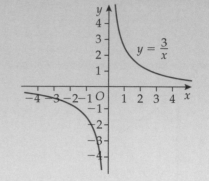

x	-3	-2	-1	1	2	3
y	-1	-1.5	-3	3	1.5	1

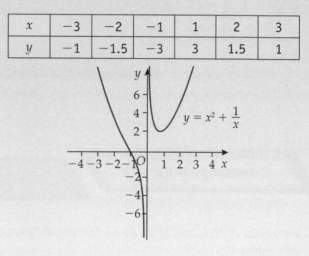

x	-2	-1	$-\frac{1}{2}$	$\frac{1}{2}$	1	2
y	3.5	0	-1.75	2.25	2	4.5

Graphs of exponential functions   A   A*

An exponential function has the form $y = k^x$, where k is a positive number. Exponential graphs can have one of two shapes, depending on the value of k.

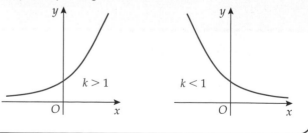

1 The diagram shows the curve $y = x^2 + 4x - 2$.

A

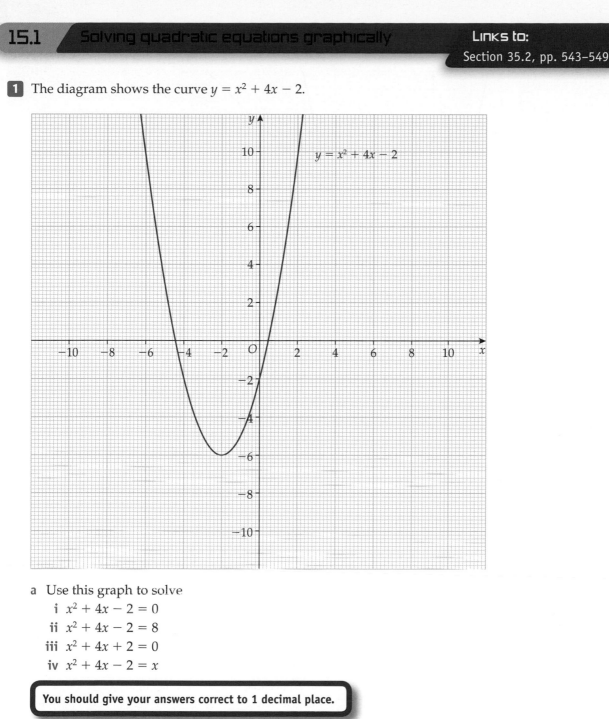

$y = x^2 + 4x - 2$

a   Use this graph to solve
   i $x^2 + 4x - 2 = 0$
   ii $x^2 + 4x - 2 = 8$
   iii $x^2 + 4x + 2 = 0$
   iv $x^2 + 4x - 2 = x$

You should give your answers correct to 1 decimal place.

b   Write down the equation of the line you could draw on this
   graph to solve the equation
   i $x^2 + 4x - 2 = 2x + 1$
   ii $x^2 + 4x - 5 = 0$
   iii $x^2 + 2x - 2 = 0$
   iv $x^2 + 5x - 10 = 0$

A02

2 The diagram shows the graphs of the curve $y = x^2 - 2x + 5$ and the line $y = x + 4$.

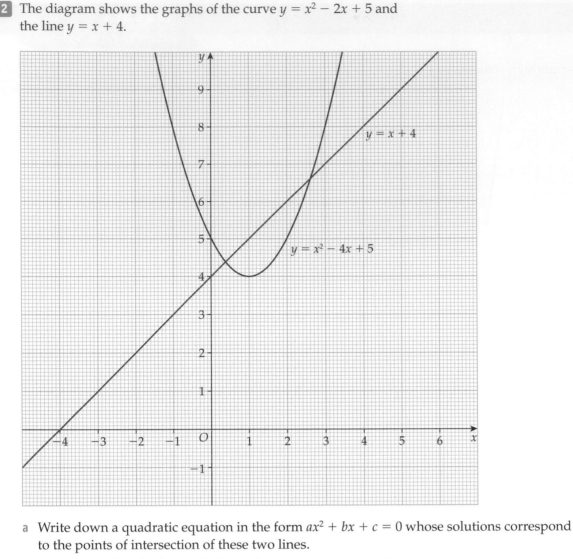

y = x + 4

y = x² - 4x + 5

a Write down a quadratic equation in the form $ax^2 + bx + c = 0$ whose solutions correspond to the points of intersection of these two lines.

b Use the graph to find the solutions of your quadratic equation.

3 Marta draws the curve $y = 2x^2 - 6x + 1 = 0$ on a graph.
Write down the equation of the line she should draw if she wants to solve the equation

a $2x^2 - 6x + 8 = 0$

b $2x^2 - 5x + 2 = 0$

c $2x^2 - 3x - 4 = 0$

d $2x^2 - 9x + 1 = 0$

4 a Draw the graph of $y = x^2 + 3x - 4$ for values of x between -5 and 2.

b By drawing an appropriate line, solve the equation $x^2 + 2x - 7 = 0$.

5 Use the discriminant to show that the lines with equations $y = x^2 + 4x + 1$ and $y = x - 2$ do not cross.

6 This graph shows the curve $y = x^2 + x + 1$ and the curve $y = 4x^2 - 3x - 3$.

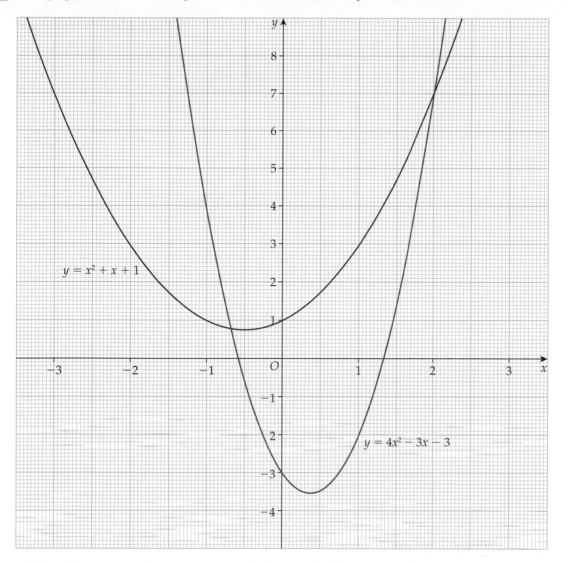

$y = x^2 + x + 1$

$y = 4x^2 - 3x - 3$

a Write down a quadratic equation in the form $ax^2 + bx + c = 0$ whose solutions correspond to the points of intersection of these two curves.

b Use the graph to find the solutions of your quadratic equation.

AO3

Challenge yourself

You won't encounter questions like this in the exam, but the underlying maths is covered in your GCSE course. **Have a go!**

The line $y = x + k$ is a tangent to the curve $y = x^2 + 2x - 1$.
Find the value of k.

A tangent touches the curve in one place.

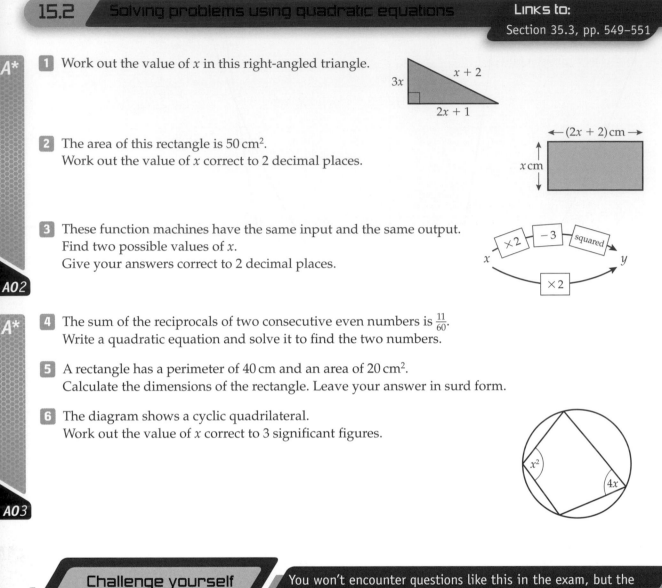

A*

1 Work out the value of x in this right-angled triangle.

2 The area of this rectangle is $50\,\text{cm}^2$.
Work out the value of x correct to 2 decimal places.

3 These function machines have the same input and the same output.
Find two possible values of x.
Give your answers correct to 2 decimal places.

AO2

A*

4 The sum of the reciprocals of two consecutive even numbers is $\frac{11}{60}$.
Write a quadratic equation and solve it to find the two numbers.

5 A rectangle has a perimeter of $40\,\text{cm}$ and an area of $20\,\text{cm}^2$.
Calculate the dimensions of the rectangle. Leave your answer in surd form.

6 The diagram shows a cyclic quadrilateral.
Work out the value of x correct to 3 significant figures.

AO3

Challenge yourself

You won't encounter questions like this in the exam, but the underlying maths is covered in your GCSE course. **Have a go!**

Martin throws a basketball from a height of two metres. The graph shows the path of the basketball. All the measurements are given in metres.

The path taken by the basketball is given by the equation
$$y = 2 + x\sqrt{3} - \frac{20x^2}{u^2}$$
where u is the initial speed of the basketball in metres per second.

a Calculate the initial speed at which Martin would have to throw the basketball in order to score.

b Martin throws the basketball at an initial speed of $8\,\text{m/s}$.
Calculate the distance along the ground from Martin to the point where the basketball first bounces.

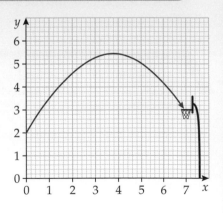

1 a Copy and complete this table of values for $y = \frac{4}{x}$.

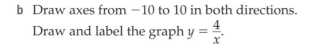

x	-8	-5	-4	-2	-1	-0.8	-0.5	-0.4	0.4	0.5	0.8	1	2	4	5	8
y	-0.5															

b Draw axes from -10 to 10 in both directions.
Draw and label the graph $y = \frac{4}{x}$.

2 a Use a table of values to draw the graph of $y = 2 + \frac{5}{x}$.

b Draw the line $y = x$ on the same set of axes.

c Use your answers to parts **a** and **b** to solve the equation $2 + \frac{5}{x} - x = 0$.

3 a Draw the graph of $y = \frac{5}{x - 2}$ for values of x from -10 to 10.

b Write down the equations of the asymptotes of your graph.

> An asymptote is a line which the graph gets closer to but never meets.

4 a Draw the graph of $y = \frac{-2}{x}$ for values of x from -8 to 8.

b Describe a transformation that would map the graph of $y = \frac{2}{x}$ onto the graph of $y = \frac{-2}{x}$.

5 This diagram shows parts of four reciprocal graphs.
No scale has been given on the axes.

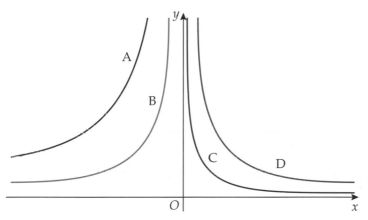

Match each equation to the correct graph.

a $y = \frac{1}{3x}$ **b** $y = \frac{2}{x}$ **c** $y = \frac{-3}{x}$ **d** $y = \frac{-6}{x}$

6 a Copy and complete this table of values for $y = \frac{1}{2}x + \frac{5}{x}$.

x	-5	-4	-2	-1.5	-1	-0.8	0.8	1	1.5	2	4	5
$\frac{1}{2}x$			-1									
$\frac{5}{x}$			-2.5									
y			-3.5									

b Draw axes from -10 to 10 in both directions.
Draw and label the graph $y = \frac{1}{2}x + \frac{5}{x}$.

A

A

A02

*A

7 The diagram shows the graphs of $y = 0.8x - \dfrac{1}{x}$ and $y = -0.5x + 1$.

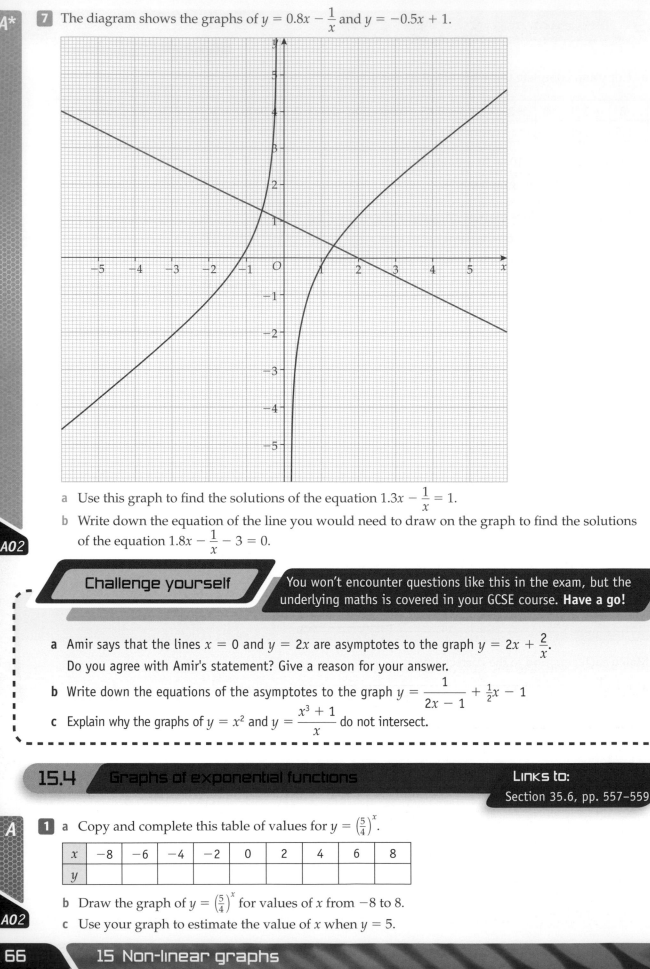

a Use this graph to find the solutions of the equation $1.3x - \dfrac{1}{x} = 1$.

b Write down the equation of the line you would need to draw on the graph to find the solutions of the equation $1.8x - \dfrac{1}{x} - 3 = 0$.

Challenge yourself

You won't encounter questions like this in the exam, but the underlying maths is covered in your GCSE course. **Have a go!**

a Amir says that the lines $x = 0$ and $y = 2x$ are asymptotes to the graph $y = 2x + \dfrac{2}{x}$.
Do you agree with Amir's statement? Give a reason for your answer.

b Write down the equations of the asymptotes to the graph $y = \dfrac{1}{2x - 1} + \dfrac{1}{2}x - 1$

c Explain why the graphs of $y = x^2$ and $y = \dfrac{x^3 + 1}{x}$ do not intersect.

15.4 **Graphs of exponential functions**

Links to:

Section 35.6, pp. 557–559

1 a Copy and complete this table of values for $y = \left(\dfrac{5}{4}\right)^x$.

x	−8	−6	−4	−2	0	2	4	6	8
y									

b Draw the graph of $y = \left(\dfrac{5}{4}\right)^x$ for values of x from −8 to 8.

c Use your graph to estimate the value of x when $y = 5$.

2 a Sketch the graph of $y = 1^x$.

b Write down the solution to the equation $4^x = 2^x$.

When you sketch a graph you don't need to use graph paper. Draw your axes with a ruler and label any points where the graph crosses the axes.

3 A biologist observing a yeast culture notices that it grows by 8% every hour.
At the beginning of the experiment the yeast culture had a mass of 4.2 grams.

a Find a formula for the mass M of the yeast culture T hours from the beginning of the experiment.

b Estimate the mass of the yeast culture 5 hours and 20 minutes after the beginning of the experiment.

4 This graph shows the exponential curve $y = \left(\frac{1}{2}\right)^x$ and the line $y = 3 - x$.
Use this graph to solve the equation $\left(\frac{1}{2}\right)^x + x - 3 = 0$.

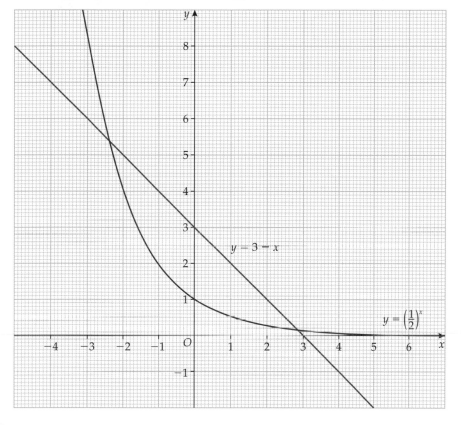

5 a Copy and complete this table of values for $y = 0.8^x$.

x	-8	-6	-4	-2	0	2	4	6	8
y									

b Draw the graph of $y = 0.8^x$ for values of x from -8 to 8.

c By drawing another line on your graph, solve the equation $0.8^x - x - 4 = 0$.

6 A small fishing lake is stocked with trout.
The number of trout in the lake T weeks after the stocking date is modelled by the formula
$$N = 80 \times 0.8^T$$

a How many trout were put into the lake on the stocking date?

b Estimate the number of trout remaining after 2 weeks.

c Draw a graph to show the number of trout in the lake for the first 15 weeks.

d Use your graph to estimate how long it will take for the number of trout in the lake to reduce by 75%.

Challenge yourself

You won't encounter questions like this in the exam, but the underlying maths is covered in your GCSE course. **Have a go!**

The game of chess was invented by an Indian mathematician. He was rewarded by the Emperor, who offered to place one grain of rice on the first square of his chessboard, two on the second, four on the third, eight on the fourth, and so on. On each square of the chessboard, the number of grains of rice would be doubled.

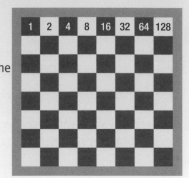

a Write the number of grains of rice on the final square of the chessboard as a power of 2.

b If this number was written out in full, how many digits would it contain?

You can use your calculator to help you with part b.

c Copy and complete this table showing the total number of grains of rice on the chessboard after each square has been filled.

Number of squares filled	1	2	3	4	5	6	7	8
Number of grains on board	1	3	7					

d Use your answer to part **c** to make a conjecture about the value of the sum

$$2^0 + 2^1 + 2^2 + 2^3 + ... + 2^{n-1}$$

A conjecture is something you think might be true but haven't proved yet.

e Prove your conjecture.

f Write an expression for the total number of grains of rice on the chessboard after all 64 squares have been filled.

Key Points

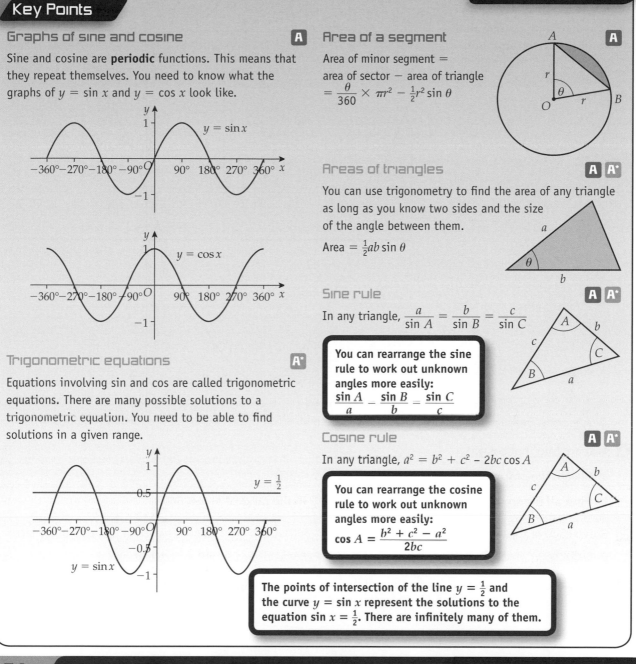

Graphs of sine and cosine **A**

Sine and cosine are **periodic** functions. This means that they repeat themselves. You need to know what the graphs of $y = \sin x$ and $y = \cos x$ look like.

Trigonometric equations **A***

Equations involving sin and cos are called trigonometric equations. There are many possible solutions to a trigonometric equation. You need to be able to find solutions in a given range.

The points of intersection of the line $y = \frac{1}{2}$ and the curve $y = \sin x$ represent the solutions to the equation $\sin x = \frac{1}{2}$. There are infinitely many of them.

Area of a segment **A**

Area of minor segment =
area of sector − area of triangle
$= \dfrac{\theta}{360} \times \pi r^2 - \dfrac{1}{2}r^2 \sin \theta$

Areas of triangles **A** **A***

You can use trigonometry to find the area of any triangle as long as you know two sides and the size of the angle between them.

Area $= \frac{1}{2}ab \sin \theta$

Sine rule **A** **A***

In any triangle, $\dfrac{a}{\sin A} = \dfrac{b}{\sin B} = \dfrac{c}{\sin C}$

You can rearrange the sine rule to work out unknown angles more easily:
$\dfrac{\sin A}{a} = \dfrac{\sin B}{b} = \dfrac{\sin C}{c}$

Cosine rule **A** **A***

In any triangle, $a^2 = b^2 + c^2 - 2bc \cos A$

You can rearrange the cosine rule to work out unknown angles more easily:
$\cos A = \dfrac{b^2 + c^2 - a^2}{2bc}$

16.1 Solving trigonometric equations

Links to:
Sections 36.1, 36.2,
pp. 566–573

1 a Copy and complete this table of values for $y = \cos x$. **A**

x	0°	5°	10°	15°	20°	25°	30°	35°	40°	50°	60°	70°	80°	90°
$\cos x$	1	1.0	0.98											

b Draw axes from 0 to 90° in the horizontal direction and from 0 to 1 in the vertical direction. Draw the graph of $y = \cos x$ for the range $0 \leqslant x \leqslant 90°$.

c Draw the line $y = 0.8$ on your graph. Use your line to find a solution to the equation $\cos x = 0.8$ in the range $0 \leqslant x \leqslant 90°$. Give your answer correct to the nearest degree.

d Use your calculator to find $\cos^{-1}(0.8)$. How accurate was your answer to part **c**?

2 This diagram shows part of the graph of $y = \sin x$ and the line $y = -\frac{1}{4}$.

Use the graph to find all the solutions to the equation $\sin x = -\frac{1}{4}$ in the range $-180° \leqslant x \leqslant 0$.

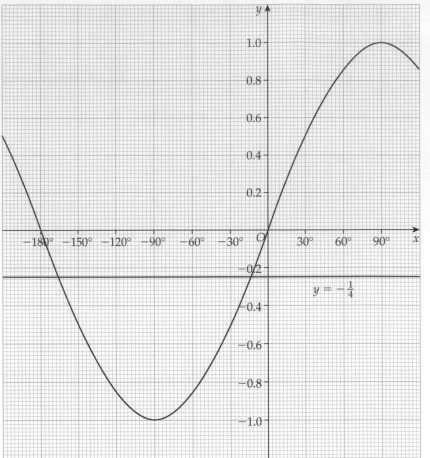

3 You are told that $\cos 45° = \dfrac{1}{\sqrt{2}}$.

a Sketch the graph of $y = \cos x$ in the range $0 \leqslant x \leqslant 360°$.

b Find all the solutions to the equation $\cos x = \dfrac{1}{\sqrt{2}}$ in the range $0 \leqslant x \leqslant 360°$.

4 Solve each of these equations in the range $0 \leqslant x \leqslant 360°$. Give your answers correct to 1 decimal place.

a $\sin x = \frac{1}{10}$ b $\cos x = 0.6$ c $\cos x = -\frac{2}{3}$ d $\sin x = -0.75$

5 Briony is creating a system for finding all the solutions to a trigonometric equation. Unfortunately, she has spilt ink on her notebook.

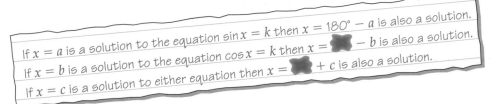

If $x = a$ is a solution to the equation $\sin x = k$ then $x = 180° - a$ is also a solution.
If $x = b$ is a solution to the equation $\cos x = k$ then $x = $ ▓ $ - b$ is also a solution.
If $x = c$ is a solution to either equation then $x = $ ▓ $ + c$ is also a solution.

a Copy and complete Briony's rules for finding solutions to trigonometric equations.

b One solution to the equation $\cos x = 0.9$ is $x = 25.8°$.

Use Briony's rules to find all the solutions in the range $0 \leqslant x \leqslant 720°$.

6 a Sketch the graph of $y = \sin x$ in the range $0 \leqslant x \leqslant 360°$.

b Find all the solutions of the equation $(\sin x)^2 = 0.36$ in the range $0 \leqslant x \leqslant 360°$. Give your answers correct to 1 decimal place.

7 Solve the equation $2 - 5 \sin x = 0$ in the range $-180° \leqslant x \leqslant 180°$.

Challenge yourself

You won't encounter questions like this in the exam, but the underlying maths is covered in your GCSE course. **Have a go!**

Find all the solutions to the equation $\cos(2x + 60°) = 0.7$ in the range $0 \leqslant x \leqslant 360°$.

There are four solutions.

16.2   Calculating areas using trigonometry

Links to:
Section 36.3, pp. 573 578

1 Calculate the areas of these triangles correct to 3 significant figures.

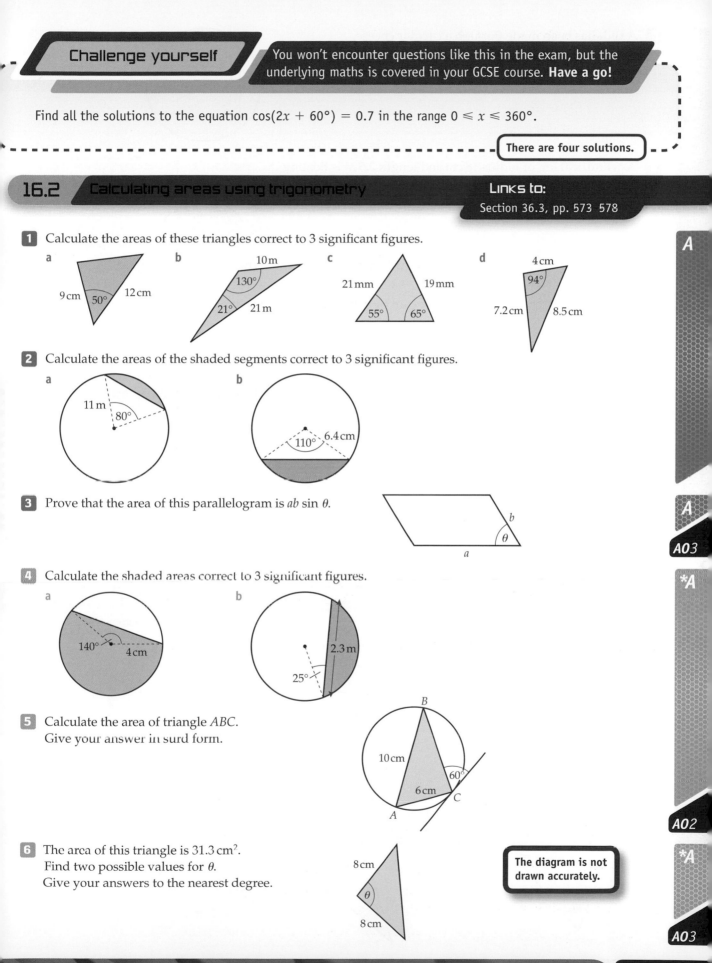

a
9 cm, 50°, 12 cm

b
10 m, 130°, 21°, 21 m

c
21 mm, 19 mm, 55°, 65°

d
4 cm, 94°, 7.2 cm, 8.5 cm

A

2 Calculate the areas of the shaded segments correct to 3 significant figures.

a
11 m, 80°

b
110°, 6.4 cm

3 Prove that the area of this parallelogram is $ab \sin \theta$.

b, θ, a

A

AO3

4 Calculate the shaded areas correct to 3 significant figures.

*A

a
140°, 4 cm

b
2.3 m, 25°

5 Calculate the area of triangle ABC.
Give your answer in surd form.

B, 10 cm, 60°, 6 cm, C, A

AO2

6 The area of this triangle is 31.3 cm^2.
Find two possible values for θ.
Give your answers to the nearest degree.

8 cm, θ, 8 cm

The diagram is not drawn accurately.

*A

AO3

7 A slice has been made through a circular cheese.
Calculate the percentage of the cheese that remains.
Give your answer correct to 1 decimal place.

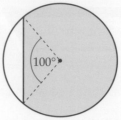

8 A cylindrical log of radius 18 cm and length 2.5 m is floating
in water.
The radius of the log makes an angle of 26° with the
water line.
Calculate

  a the volume of the submerged section of the log

  b the surface area of the log which is in contact with the air.

Give your answers correct to 3 significant figures.

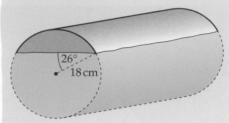

Challenge yourself

You won't encounter questions like this in the exam, but the
underlying maths is covered in your GCSE course. **Have a go!**

A regular icosahedron is a solid shape with 20 identical faces. Each face is an
equilateral triangle.
If the side length of each triangular face is n cm, find an expression for the
total surface area of the icosahedron.
Give your answer in surd form.

16.3 The sine rule

Links to:
Section 36.4, pp. 578–581

1 Calculate the length of each side marked with a letter.
Give your answers correct to 3 significant figures.

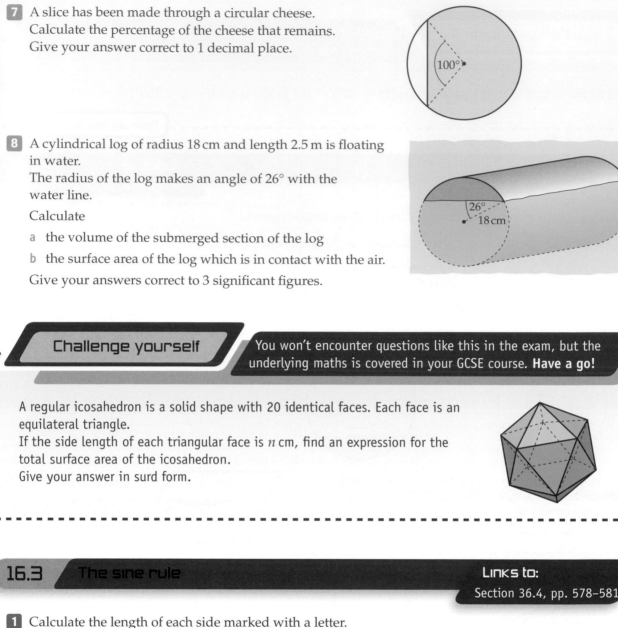

a

70°   7.5 cm   50° a

b

b 70°   80°   11 m

c

40 m   26°   9° c

2 Calculate the size of each angle marked with a letter.
Give your answers correct to 1 decimal place.

a

9 cm a 8 cm   60°

b

9.5 m   45°   12.1 m b

c

c 15 mm   30°   27 mm

3 Work out the area of this triangle.

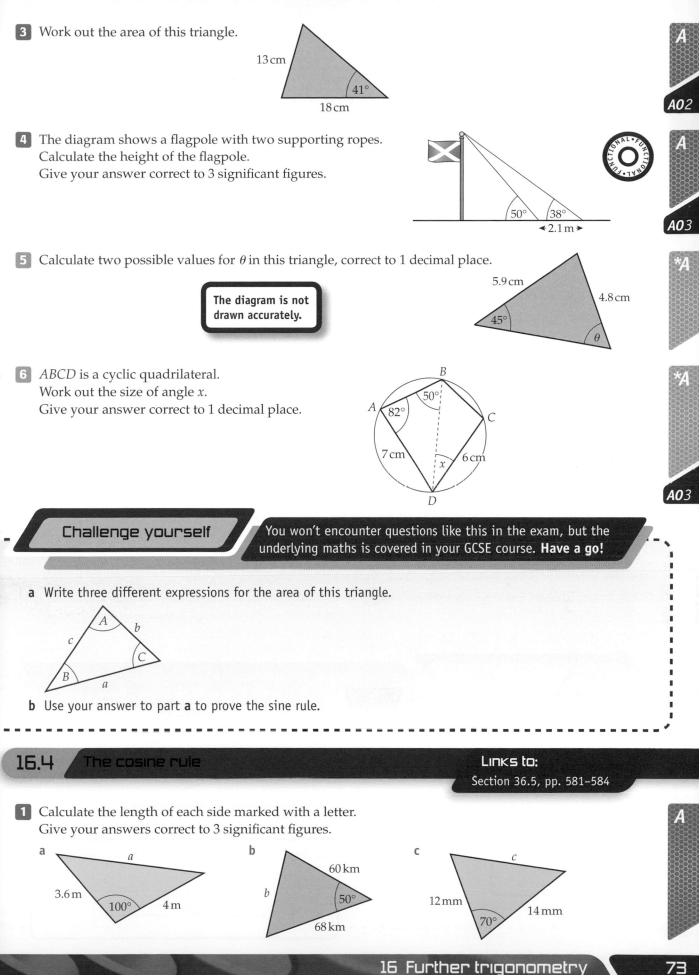

13 cm
41°
18 cm

4 The diagram shows a flagpole with two supporting ropes.
Calculate the height of the flagpole.
Give your answer correct to 3 significant figures.

50° 38°
◄ 2.1 m ►

5 Calculate two possible values for θ in this triangle, correct to 1 decimal place.

The diagram is not drawn accurately.

5.9 cm
4.8 cm
45°
θ

6 *ABCD* is a cyclic quadrilateral.
Work out the size of angle *x*.
Give your answer correct to 1 decimal place.

B
50°
A 82°
C
7 cm
x 6 cm
D

Challenge yourself

You won't encounter questions like this in the exam, but the underlying maths is covered in your GCSE course. **Have a go!**

a Write three different expressions for the area of this triangle.

A
c b
B C
a

b Use your answer to part **a** to prove the sine rule.

16.4 The cosine rule

Links to:
Section 36.5, pp. 581–584

1 Calculate the length of each side marked with a letter.
Give your answers correct to 3 significant figures.

a
a
3.6 m
100° 4 m

b
60 km
b 50°
68 km

c
c
12 mm
70° 14 mm

A
AO2
A
AO3
*A
*A
AO3
A

2 Calculate the size of each angle marked with a letter.
Give your answers correct to 1 decimal place.

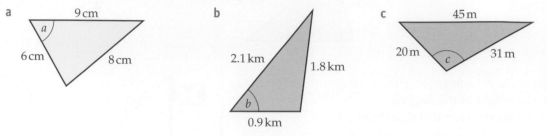

a 9 cm 6 cm 8 cm *a*

b 2.1 km 1.8 km 0.9 km *b*

c 45 m 20 m 31 m *c*

3 Freddie is using compasses to draw a circle.
The arms of his compasses are 9 cm long and he opens them to an angle of 14°.
Calculate the area of his circle correct to 3 significant figures.

14° 9 cm

4 A triangle has sides of length 3 cm, 7 cm and 8 cm. Show that it contains an angle of 60°.

5 A canoeist and a swimmer leave a jetty at the same time.
The canoeist travels on a bearing of 065° at a constant speed of 2.4 m/s.
The swimmer travels on a bearing of 350° at a constant speed of 0.8 m/s.
Calculate the distance between the swimmer and the canoeist after 1 hour.
Give your answer in km correct to 3 significant figures.

6 Work out the length *DC* correct to 3 significant figures.

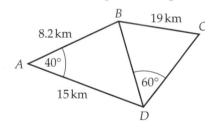

B 19 km *C*
8.2 km
A 40°
60°
15 km
D

Challenge yourself

You won't encounter questions like this in the exam, but the underlying maths is covered in your GCSE course. **Have a go!**

In this diagram $AB = BC = CD = DE = 1$ m.
Prove that $\angle AEB + \angle ADB = \angle ACB$.

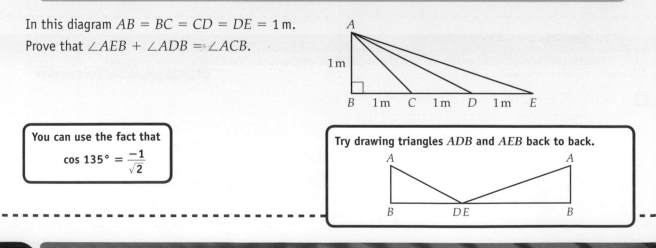

You can use the fact that
$$\cos 135° = \frac{-1}{\sqrt{2}}$$

Try drawing triangles *ADB* and *AEB* back to back.

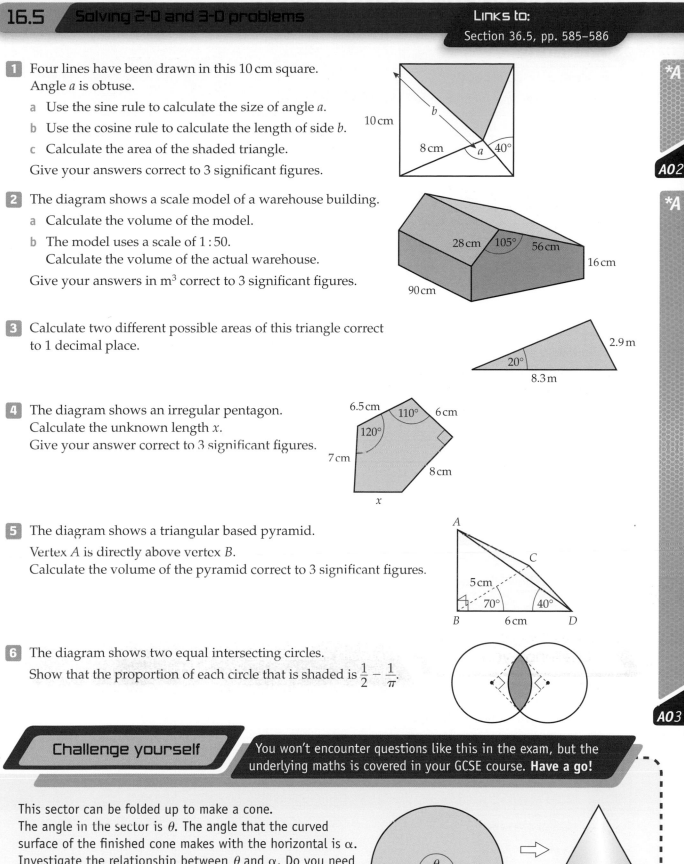

1 Four lines have been drawn in this 10 cm square.
Angle a is obtuse.

   a Use the sine rule to calculate the size of angle a.

   b Use the cosine rule to calculate the length of side b.

   c Calculate the area of the shaded triangle.

Give your answers correct to 3 significant figures.

*A

AO2

2 The diagram shows a scale model of a warehouse building.

   a Calculate the volume of the model.

   b The model uses a scale of $1:50$.
   Calculate the volume of the actual warehouse.

Give your answers in m³ correct to 3 significant figures.

*A

3 Calculate two different possible areas of this triangle correct
to 1 decimal place.

4 The diagram shows an irregular pentagon.
Calculate the unknown length x.
Give your answer correct to 3 significant figures.

5 The diagram shows a triangular based pyramid.

Vertex A is directly above vertex B.
Calculate the volume of the pyramid correct to 3 significant figures.

6 The diagram shows two equal intersecting circles.
Show that the proportion of each circle that is shaded is $\frac{1}{2} - \frac{1}{\pi}$.

AO3

Challenge yourself

You won't encounter questions like this in the exam, but the
underlying maths is covered in your GCSE course. **Have a go!**

This sector can be folded up to make a cone.
The angle in the sector is θ. The angle that the curved
surface of the finished cone makes with the horizontal is α.
Investigate the relationship between θ and α. Do you need
to know the radius of the sector to calculate α?
Can you find a formula for α in terms of θ?

Links to:
Higher Student Book
Ch37, pp. 590–598

Key Points

Transformations of graphs

A*

You can use function notation to describe transformations of graphs.

Function	$y = f(x) + a$	$y = f(x + a)$	$y = af(x)$	$y = f(ax)$
Transformation of graph	Translation $\begin{pmatrix} 0 \\ a \end{pmatrix}$	Translation $\begin{pmatrix} -a \\ 0 \end{pmatrix}$	Stretch in the vertical direction, scale factor a	Stretch in the horizontal direction, scale factor $\dfrac{1}{a}$
Example $(a > 1)$	$y = f(x) + a$ $y = f(x)$	$y = f(x + a)$ $y = f(x)$	$y = af(x)$ $y = f(x)$	$y = f(ax)$ $y = f(x)$

Transformations of trigonometric graphs

A*

The **amplitude** of a trigonometric function is the height of each peak from the centre point.
The **period** of a trigonometric function is the distance between peaks.

Period

Amplitude

17.1 Transformations of graphs

Links to:
Section 36.4, pp. 590–595

A*

1 The diagram shows the curve with equation $y = \frac{1}{2}x^2$.

a Copy this graph on to graph paper.

b Describe the translation which transforms $y = \frac{1}{2}x^2$ into $y = \frac{1}{2}(x + 4)^2$

c Draw the curve with equation $y = \frac{1}{2}(x + 4)^2$ on your graph.

d Use transformations to draw the curve with equation $y = \frac{1}{2}(x - 2)^2 - 6$ on your graph.

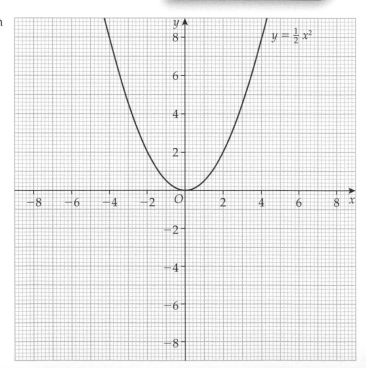

$y = \frac{1}{2}x^2$

2 The function f is defined by $f(x) = \frac{1}{2}x + 2$.
On graph paper draw axes from -10 to 10 in both directions.
Draw the line $y = f(x)$.
For each of the following lines:
  i draw the line on your graph
 ii write down the equation of the line in terms of x
iii describe the transformation that maps $y = f(x)$ onto the line.

a $f(2x)$

b $f(x - 5)$

c $3f(x)$

d $\frac{1}{2}f(x)$

3 Describe the transformation that maps $y = x^3$ onto

a $y = (2x)^3$

b $y = (x - 6)^3$

c $y = (x + 1)^3 - 5$

d $y = 6x^3$

4 The diagram shows a sketch of the graph of $y = f(x)$.

On separate axes, sketch graphs of the following functions.
Label the new positions of points A, B and C.

a $y = f(x) + 3$

b $y = f(x) - 2$

c $y = f(x + 1)$

d $y = f(x - 5)$

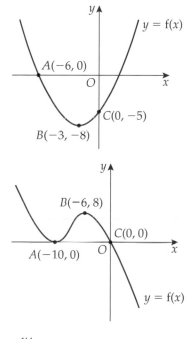

5 The diagram shows a sketch of the graph of $y = f(x)$.

On separate axes, sketch graphs of the following functions.
Label the new positions of points A, B and C.

a $y = 3f(x)$

b $y = \frac{1}{2}f(x)$

c $y = f(5x)$

d $y = f(\frac{1}{2}x)$

6 The diagram shows a sketch of the graph of $y = f(x)$.

On separate axes, sketch graphs of the following functions.
Label the new positions of points A, B and C.

a $y = f(x + 2) - 2$

b $y = 3f(x) + 5$

c $y = 2f(x - 1)$

d $y = f(2x) - 1$

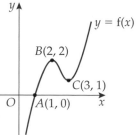

7 The graph of $y = \sqrt{3x + 1}$ is transformed by **A**, then by **B**.

A: Translation by the column vector $\begin{pmatrix} 2 \\ -1 \end{pmatrix}$

B: Stretch in the vertical direction, scale factor 4

Write down the equation of the transformed graph.

The functions f and g are defined as follows:

$$f(x) = x^2 - 5 \qquad g(x) = 2x + 3$$

a Solve the equation $f(x) + g(x^2) = 0$.

b Find an expression for $f(g(x))$.

c Work out the roots of the equation $f(g(x)) = g(f(x))$. Give your answers correct to 3 significant figures.

17.2 Transformations of trigonometric graphs

Links to:

Section 37.3, pp. 595–598

A*

1 The diagram shows the graph of $y = \sin x$.

    a Copy this graph on to graph paper.

    b Use transformations to draw the curve $y = \sin(2x) - 1$ on your graph.

2 The diagram shows a graph in the form $y = a \sin x$.

    Write down the value of a.

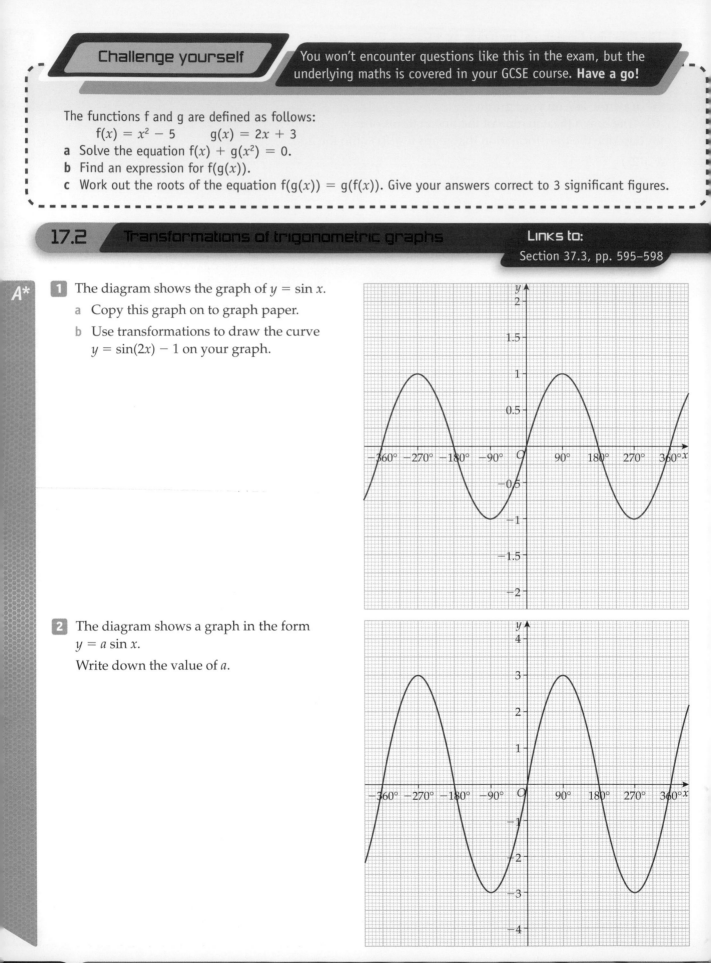

3 Describe fully the transformation which maps the graph of
$y = \sin x$ onto the graph of $y = \sin(x + 30°) + 2$.

*A

4 The diagram shows a graph in the form
$y = a\cos(x + b)$.

Work out the values of a and b.

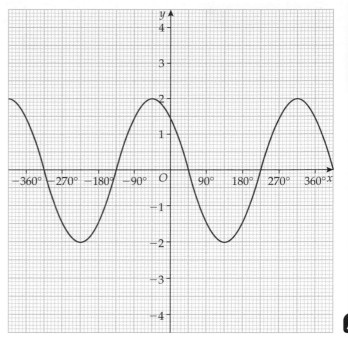

AO2

5 **a** Describe a transformation which maps the graph of $y = \sin x$ onto the graph of $y = \cos x$.

b If $\cos x = \sin(x + k)$, find the value of k.

*A

6 Choose the correct graph for each of
these equations.

a $y = 2\cos x$

b $y = \cos x - 3$

c $y = \cos(x - 45°)$

d $y = \cos 4x$

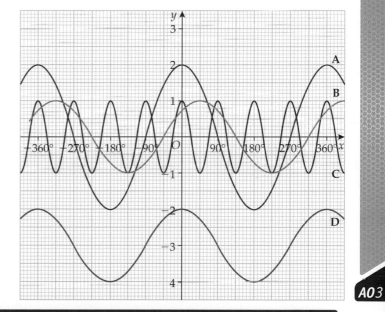

7 Work out the period and amplitude of
each of the following graphs.

a $y = 2\sin x$

b $y = \cos(x + 45°)$

c $y = \sin(3x)$

d $y = \cos(6x) - 2$

AO3

Challenge yourself

You won't encounter questions like this in the exam, but the
underlying maths is covered in your GCSE course. **Have a go!**

The depth of the water in a river estuary is modelled using the equation $d = 12 + 8\sin(30t)$
where d is the depth of the water in feet and t is the time after midnight in hours.
a What is the minimum depth of the water in the estuary?
b At what time during the day does this minimum value occur?
c When the water depth is below 8.5 feet, a causeway allows tourists to visit an island in the estuary.
For how long each day is the island accessible? Give your answer correct to 2 decimal places.

Key Points

Vectors

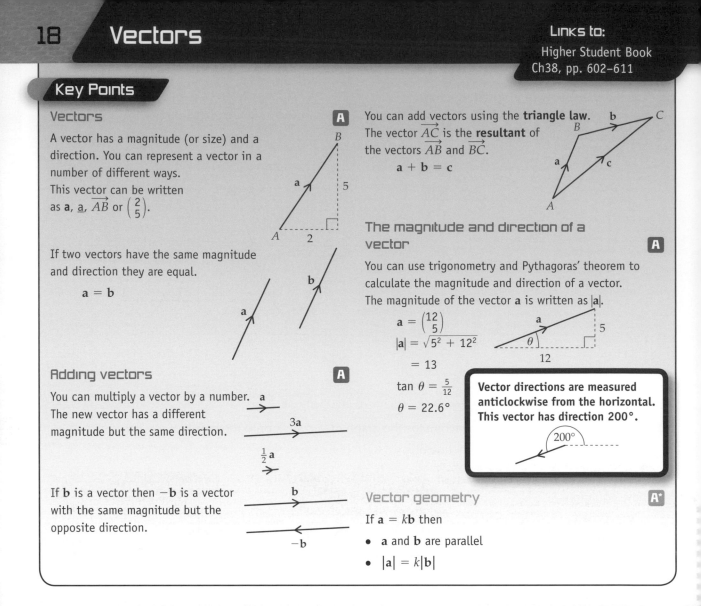

A vector has a magnitude (or size) and a direction. You can represent a vector in a number of different ways.
This vector can be written as **a**, <u>a</u>, $\overrightarrow{AB}$ or $\begin{pmatrix} 2 \\ 5 \end{pmatrix}$.

If two vectors have the same magnitude and direction they are equal.

$$\mathbf{a} = \mathbf{b}$$

Adding vectors

You can multiply a vector by a number. The new vector has a different magnitude but the same direction.

If **b** is a vector then −**b** is a vector with the same magnitude but the opposite direction.

You can add vectors using the **triangle law**.
The vector $\overrightarrow{AC}$ is the **resultant** of the vectors $\overrightarrow{AB}$ and $\overrightarrow{BC}$.

$$\mathbf{a} + \mathbf{b} = \mathbf{c}$$

The magnitude and direction of a vector

You can use trigonometry and Pythagoras' theorem to calculate the magnitude and direction of a vector.
The magnitude of the vector **a** is written as |**a**|.

$$\mathbf{a} = \begin{pmatrix} 12 \\ 5 \end{pmatrix}$$
$$|\mathbf{a}| = \sqrt{5^2 + 12^2}$$
$$= 13$$
$$\tan \theta = \frac{5}{12}$$
$$\theta = 22.6°$$

Vector directions are measured anticlockwise from the horizontal. This vector has direction 200°.

Vector geometry

If $\mathbf{a} = k\mathbf{b}$ then

- **a** and **b** are parallel
- $|\mathbf{a}| = k|\mathbf{b}|$

18.1 Vectors

Links to:
Sections 38.1, 38.2
pp. 602–608

A

1 The vectors $\mathbf{a} = \begin{pmatrix} 2 \\ 3 \end{pmatrix}$ and $\mathbf{b} = \begin{pmatrix} 4 \\ -1 \end{pmatrix}$ are shown on this grid.

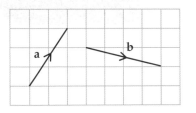

Draw these vectors on squared paper.

  a 2**a**   **b** $\frac{1}{2}$**b**   **c** −**a**

  d **a** + **b**   **e** **b** − **a**   **f** 2**a** − **b**

2 $\mathbf{r} = \begin{pmatrix} 1 \\ 1 \end{pmatrix}$ $\mathbf{s} = \begin{pmatrix} -5 \\ 2 \end{pmatrix}$ $\mathbf{t} = \begin{pmatrix} -2 \\ -1 \end{pmatrix}$

Write each of these vectors as a column vector.

  a 10**t**   **b** −**r**   **c** **r** + **s**

  d 2**r** − **t**   **e** **r** − 2**t**   **f** **r** + **s** − **t**

3 The vector **a** has magnitude 6 and direction 18°.

Write down the magnitude and direction of these vectors.

    a 2**a**   **b** $\frac{1}{3}$**a**   **c** −**a**

4 Work out the magnitude and direction of these vectors.

    a $\begin{pmatrix} 3 \\ 4 \end{pmatrix}$ **b** $\begin{pmatrix} -2 \\ 4 \end{pmatrix}$ **c** $\begin{pmatrix} -6 \\ 6 \end{pmatrix}$ **d** $\begin{pmatrix} -1 \\ -7 \end{pmatrix}$

5 On this grid, $\overrightarrow{AB} = $ **a** and $\overrightarrow{AD} = $ **b**.

Write these vectors in terms of **a** and **b**.

    a $\overrightarrow{AC}$ **b** $\overrightarrow{AE}$ **c** $\overrightarrow{DA}$

    d $\overrightarrow{AH}$ **e** $\overrightarrow{HC}$ **f** $\overrightarrow{ID}$

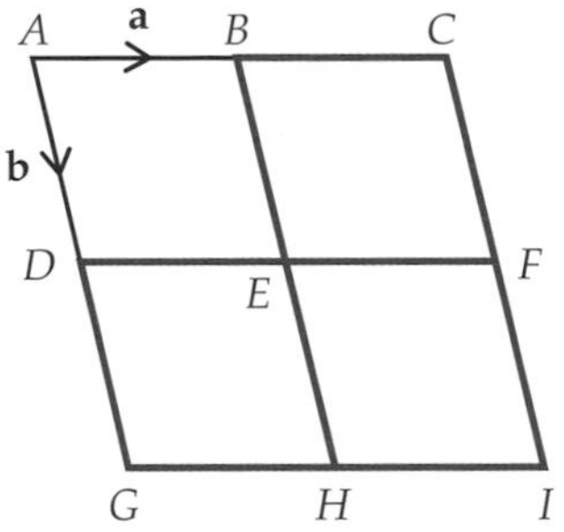

6 The diagram shows a grid of equilateral triangles.

$\overrightarrow{OP} = $ **a** and $\overrightarrow{OR} = $ **b**.

Write these vectors in terms of **a** and **b**.

    a $\overrightarrow{RS}$ **b** $\overrightarrow{SP}$ **c** $\overrightarrow{OS}$

    d $\overrightarrow{RP}$ **e** $\overrightarrow{QT}$ **f** $\overrightarrow{QR}$

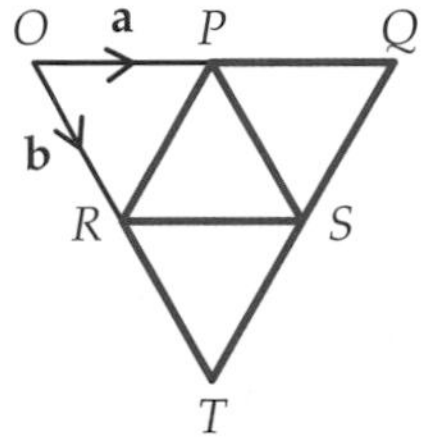

Challenge yourself

You won't encounter questions like this in the exam, but the underlying maths is covered in your GCSE course. **Have a go!**

Investigate the area of the parallelogram with sides **a** = $\begin{pmatrix} p \\ q \end{pmatrix}$ and **b** = $\begin{pmatrix} r \\ s \end{pmatrix}$.

Can you prove that the parallelogram has area $|ps - rq|$?

The vertical lines mean it is the *size* of your answer that matters, not whether it is positive or negative.

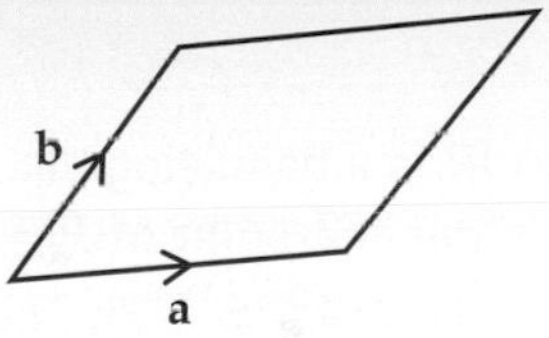

18.2   Vector geometry

Links to:

Section 38.3, pp. 608–611

1 *OPQR* is a parallelogram.

Point *M* lies on the line *OR* such that $OM:MR = 4:1$.

Point *N* is the mid-point of *RQ*. $\overrightarrow{OP} = $ **a** and $\overrightarrow{OM} = $ **b**.

Write expressions for these vectors in terms of **a** and **b**.

    a $\overrightarrow{OR}$ **b** $\overrightarrow{RN}$ **c** $\overrightarrow{MP}$

    d $\overrightarrow{NP}$ **e** $\overrightarrow{NO}$ **f** $\overrightarrow{MN}$

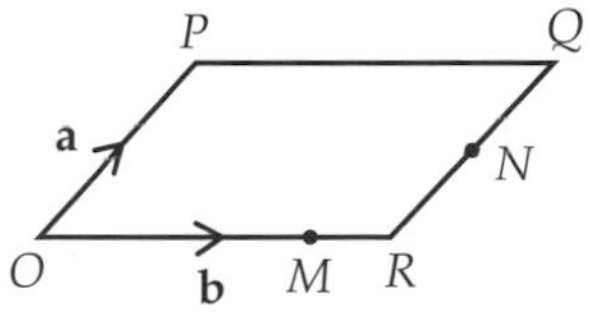

2 *OPQ* is a triangle.

$OM:MP = ON:NQ = 1:2$. $\overrightarrow{OM} = $ **a** and $\overrightarrow{ON} = $ **b**.

    a Write expressions for $\overrightarrow{MN}$ and $\overrightarrow{PQ}$ in terms of **a** and **b**.

    b What does your answer to part **a** tell you about the lines *MN* and *PQ*?

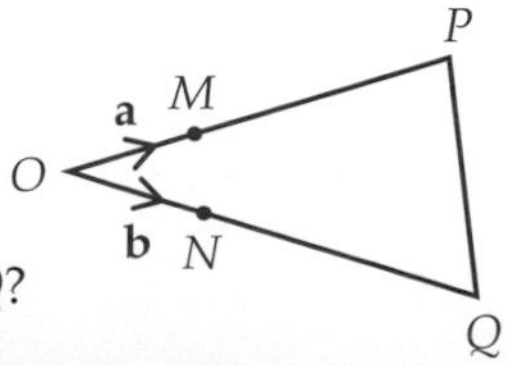

3 The triangle PQR is formed by joining the mid-points of the sides of the triangle ABC.

Use vectors to prove that the two triangles are similar.

4 The quadrilateral $PQRS$ is formed by joining the mid-points of the sides of the parallelogram $ABCD$.

Use vectors to prove that $PQRS$ is also a parallelogram.

5 OPQ is a triangle.

$2\overrightarrow{PR} = \overrightarrow{RQ}$ and $3\overrightarrow{OR} = \overrightarrow{OS}$.

$\overrightarrow{OP} = \mathbf{a}$ and $\overrightarrow{OQ} = \mathbf{b}$.

a Show that $\overrightarrow{OS} = 2\mathbf{a} + \mathbf{b}$.

b Point T is added to the diagram such that $\overrightarrow{OT} = -\mathbf{b}$.
Prove that points T, P and S lie on a straight line.

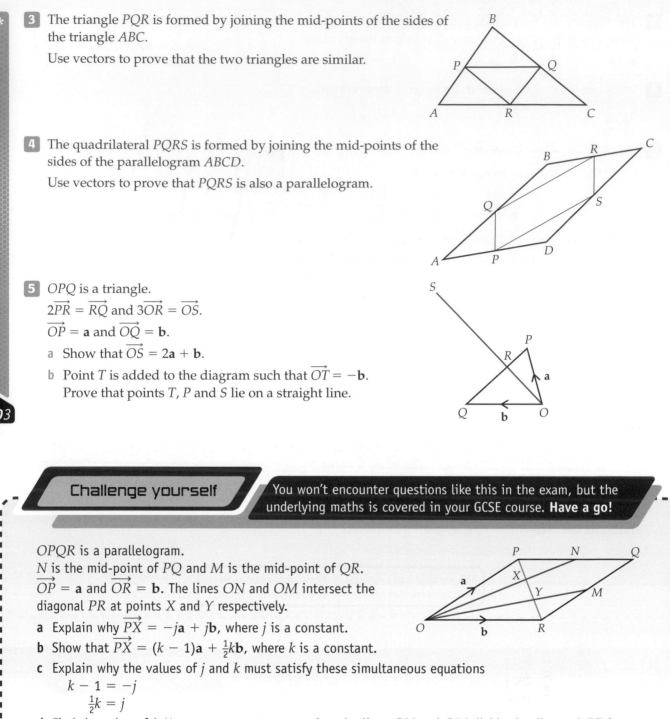

Challenge yourself

You won't encounter questions like this in the exam, but the underlying maths is covered in your GCSE course. **Have a go!**

$OPQR$ is a parallelogram.

N is the mid-point of PQ and M is the mid-point of QR.

$\overrightarrow{OP} = \mathbf{a}$ and $\overrightarrow{OR} = \mathbf{b}$. The lines ON and OM intersect the diagonal PR at points X and Y respectively.

a Explain why $\overrightarrow{PX} = -j\mathbf{a} + j\mathbf{b}$, where j is a constant.

b Show that $\overrightarrow{PX} = (k - 1)\mathbf{a} + \frac{1}{2}k\mathbf{b}$, where k is a constant.

c Explain why the values of j and k must satisfy these simultaneous equations

$k - 1 = -j$

$\frac{1}{2}k = j$

d Find the value of j. Use your answer to prove that the lines ON and OM divide the diagonal PR into three equal parts.

EXAM PRACTICE PAPERS

Unit 1 Higher Statistics and Number

1 hr, Calculator allowed

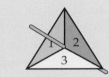

1 Indie has these two spinners.

She spins the spinners at the same time.

She subtracts the smaller number from the larger number to give her a score.

The table shows her possible scores.

−	1	2	3	4	5
1	0	1	2	3	4
2	1	0	1	2	3
3	2	1	0	1	2

    a   Work out the probability Indie gets a score of 0.   **(1 mark)** | D

    b   Work out the probability that her score is an odd number.   **(2 marks)** | D

2   A petrol strimmer costs £92 + $17\frac{1}{2}$% VAT. Work out the total cost of the strimmer.   **(2 marks)** | D | **Funct.**

3   These are two 90-day prices plans offered by an electricity supplier.

    Plan A: **£0.14 per unit of electricity used**

    Plan B: **£0.12 per unit of electricity used + £12.70 fixed charge**

In 90 days Mr Vaughan uses, on average, 1200 units of electricity.

Which plan would it be best for Mr Vaughan to use?

You must show all your working.   **(3 marks)** | D | **AO3** | **Funct.**

4   Rohan wants to find out how much exercise people do in a week.

He has written this question for his survey.

> How much exercise do you do in one week? Please tick one box only.
>
> 2 to 4 hours ☐   4 to 6 hours ☐   6 to 8 hours ☐

    a   Write down two criticisms of the question.   **(2 marks)** | D | **AO2** | **Funct.**

    b   Re-write the question to make it more suitable.   **(2 marks)** | C | **AO2** | **Funct.**

5   An estate agent collects information on the average prices of three-bedroom houses at certain distances from a motorway. The table shows his results.

Distance from motorway (km)	2	12	5	3	11	15
Average price (£000s)	164	200	176	172	196	214

    a   Draw a scatter diagram to show this information on a copy of the coordinate grid below.   **(2 marks)** D | **Funct.**

[Scatter diagram grid: Average price (£000s) on vertical axis from 160 to 220, Distance from motorway (km) on horizontal axis from 0 to 16]

    b   The estate agent says that the closer a house is to the motorway, the cheaper it is.

        Do you agree with this statement? Give a reason for your answer.   **(1 mark)** | D | **Funct.**

6* The table shows the number of take-away meals eaten each week by a sample of 60 adults.

Number of take-away meals	0	1	2	3	4	5	more than 5
Frequency	16	12	9	8	5	6	4

    a   Is it possible to calculate the mean of this data? Give a reason for your answer.   **(2 marks)** | D | **AO2**

    b   Is it possible to calculate the median of this data? Give a reason for your answer.   **(2 marks)** | D | **AO2**

7 In 2009 the number of people living in a village was 850.

In 2010 the number of people living in the village was 918.

A local councillor says

    'If the number of people continues to increase by the same percentage each year, by 2015 there will be more than 1200 people in the village.'

Do you agree with the local councillor?

You must show working to support your answer.   **(3 marks)** | C | AO3 | **Funct.**

8* Nikki and Sara are business partners. Each year they share a £12 000 bonus in the ratio of the number of years they have been in the business.

This year Nikki has been in the business 6 years and Sara has been in the business 2 years.

Show that in four years' time the difference between the amounts they receive will be halved.   **(6 marks)** | C | AO3

9 The speed of light is approximately 300 million metres per second.

  a Write this number in standard form.   **(1 mark)** | B

  b Multiply your answer in part **a** by 60.
Give your answer in standard form.   **(1 mark)** | B

  c What do you think your answer to part b represents?   **(1 mark)** | B | AO2 | **Funct.**

10 This table shows the time taken by 60 adults to complete a puzzle.

Time, t, minutes	Frequency
$0 < t \leq 1$	15
$1 < t \leq 2$	12
$2 < t \leq 3$	23
$3 < t \leq 4$	6
$4 < t \leq 5$	4

  a Draw a cumulative frequency diagram to illustrate this information.   **(3 marks)** | B | AO2

  b Use your graph to estimate the median.   **(1 mark)** | B

  c **i** Explain why your answer to part **b** is an estimate.   **(1 mark)** | B | AO2

     ii Explain what your answer to part **b** represents.   **(1 mark)** | B | AO2

  d Sue says 'Over 65% of the adults took longer than $2\frac{1}{2}$ minutes to solve the puzzle.'
Is Sue correct? Explain your answer.   **(2 marks)** | B | AO2

11 A survey was carried out at a railway station one week to find out how many trains were late, and by how many minutes (to the nearest minute). This histogram shows the results of the survey.

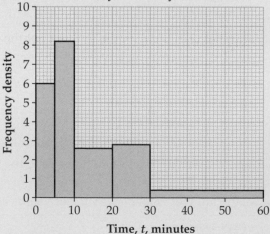

Histogram showing number of late trains and by how many minutes.

Frequency density / Time, t, minutes

  a What is the total number of trains that were late?   **(3 marks)** | A

  b How many trains were more than 20 minutes late?   **(1 mark)** | A

  c Work out an estimate of the mean number of minutes a train is late.   **(3 marks)** | A* | AO2

12 The table shows the ages of the members of a tennis club.

Age (years)	Under 15	16–30	31–45	Over 45
Number of members	42	28	16	24

The manager of the tennis club wants a stratified sample of 25 people.

How many members should be chosen from each age group? (3 marks) | A | Funct.

13 A science test is in two parts, a written test and a practical test. Out of all the people who sit the written test, 85% pass. When a person passes the written test, the probability that they pass the practical test is 70%. When a person fails the written test, the probability that they fail the practical test is 65%.

What is the probability that a person chosen at random

a fails both tests (2 marks) | A*

b passes exactly one test? (3 marks) | A*

Unit 2 Higher Number and Algebra   1 hr 15 min, Non-calculator

1 The formula to find the area, A, of a rectangle is

$A = l \times w$ where l is the length and w is the width of the rectangle.

A rectangle has a length of $2x$ cm and a width of $3x - 1$ cm.

a Show that the formula for the area, A, of the rectangle is

$A = 6x^2 - 2x$ (1 mark) | D

b Work out the value of A when $x = 6$. (2 marks) | D

2 A two-stage operation is shown.

Input ⟶ Subtract 5 ⟶ Multiply by 2 ⟶ Output

a Work out the output when the input is -3. (1 mark) | D

b When the input is n what is the output? (2 marks) | D

3 The nth term of a sequence is $4n + 5$.

Show that all the terms in the sequence are odd. (2 marks) | D | AO2

4* In the first quarter of the year Melia used 1000 units of electricity.

Each unit of electricity cost 10p

In the second quarter of the year Melia used 10% less units, but each unit cost 10% more.

Will Melia's electricity bill be the same in the second quarter of the year as in the first quarter?

You **must** show your working. (4 marks) | D | AO2 | Funct.

5 An electricity supplier uses this formula to work out the total cost of the electricity a customer uses.

$C = 0.1U + 12$ C is the total cost of the electricity in pounds

U is the number of units of electricity used

a Sam uses 850 units of electricity.

What is the total cost of the electricity she uses? (2 marks) | D | Funct.

b The total cost of the electricity Clive used is £84

How many units of electricity did he use? (2 marks) | C | AO2 | Funct.

6 Use approximations to estimate the value of

$$\frac{8105}{21.76 \times 0.219}$$ (3 marks) | C

7 The surface area of a cube is $(6x + 18)$ cm².

Three of these cubes are used to make the cuboid shown.

The surface area of the cuboid is 98 cm².

Work out the value of x. (5 marks) | C | AO3

8* Steffan works for a company as a sales representative.

At the start of 2009 he bought a car for £20 000.

At the end of 2009 he sold the car for £15 200.

His mileage for the year was 40 000 miles.

The average cost of running his car was 32p per mile.

His company pay him 40p per mile travelled.

What is Steffan's percentage profit or loss after buying, driving and selling the car?

You **must** show your working and clearly state whether Steffan has made a profit or a loss.

(6 marks) | C | AO3 | **Funct.**

9 At a school 60% of the students are girls and 40% are boys.

On one particular day, 10% of the girls and 20% of the boys have the flu.

What percentage of the pupils in the whole school have the flu on this particular day? **(3 marks)** | B | AO2

10 a Solve the equation $4x + 5 = 23 - 2x$ **(3 marks)** | D

b Solve the equation $\dfrac{x + 1}{3} + \dfrac{3x - 5}{2} = 7$ **(4 marks)** | B

11 A is the point $(3, 5)$ and B is the point $(1, -1)$.

Find the equation of the straight line parallel to AB that passes through the point $(4, 2)$.

You must show your working. **(3 marks)** | B | AO3

12 Evan completes a questionnaire.

He ticks this box to show the values his age lies between.

☑ $15 \leqslant a < 20$

a Write down all the whole number ages that Evan could be. **(1 mark)** | D

b Show the inequality $15 \leqslant a < 20$ on a copy of this number line.

12 13 14 15 16 17 18 19 20 21 22

(1 mark) | D

Evan has two brothers, Alun and Berwyn.

The sum of Alun and Berwyn's ages is 44.

The difference is 6.

c Write equations for the sum of their ages and the difference in their ages. **(1 mark)** | B

d Solve the equations simultaneously to find the ages of Alun and Berwyn. **(2 marks)** | B

13 Simplify fully $\dfrac{4x^2 - 25}{2x^2 - x - 15}$ **(5 marks)** | A

14 a Simplify fully $\dfrac{(x^3)^4}{x^2}$ **(2 marks)** | A

b Explain why $125^{-\frac{1}{3}} = \frac{1}{5}$ **(2 marks)** | A* | AO2

15 a Write the expression $x^2 - 4x - 10$ in the form $(x + a)^2 + b$. **(3 marks)** | A*

b Hence, or otherwise, solve $x^2 - 4x - 10 = 0$.

Give your answers in surd form. **(2 marks)** | A*

16 Write the expression $(3 + \sqrt{8})(7 - \sqrt{18})$ in the form $a + b\sqrt{c}$, where a, b and c are integers. **(4 marks)** | A*

Unit 3 Higher Geometry and Algebra   1 hr 30 min, Calculator allowed

1 Work out the area of this shape.

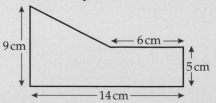

9 cm

6 cm

5 cm   Not drawn accurately

14 cm

(4 marks) | D

2 The diagram shows the position of two lighthouses, P and Q.

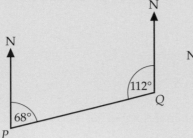

Not drawn accurately

a Explain why the bearing of P from Q is not 112° **(1 mark)** | **D** | **AO2**

b Work out the bearing of P from Q. **(2 marks)** | **D**

3 Copy this diagram.

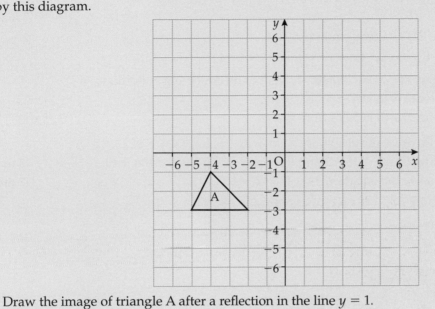

a Draw the image of triangle A after a reflection in the line $y = 1$.
Label the triangle B. **(2 marks)** | **D**

b Draw the image of triangle A after a translation by the vector $\binom{6}{4}$.
Label the triangle C. **(2 marks)** | **C**

4 a When $a = 4$ and $b = 2$, work out the value of $2a^2 - 11b$ **(2 marks)** | **D**

b Factorise $12x - 15$ **(1 mark)** | **D**

c Write down two possible values of x that would make this statement true.
$$x^2 = 16$$ **(1 mark)** | **D** | **AO2**

d Ruth compares the cost, per month, of electricity from two different companies.

SW Electricity	N Electric
£0.12 per unit **plus** £24.20 standing charge	£0.15 per unit **plus** £8.60 standing charge

i Write a formula for the total cost, £C, per month for using U units of
electricity with SW Electricity. **(1 mark)** | **D** | **AO2**

ii Write a formula for the total cost, £T, per month for using U units of
electricity with N Electric. **(1 mark)** | **D** | **AO2**

iii Ruth uses on average 600 units of electricity per month.
Which electricity company is it cheaper for her to use?
You **must** show your working. **(3 marks)** | **D** | **AO2** | **Funct.**

iv For how many units of electricity per month would the total cost from
each company be the same?
You **must** show your working. **(3 marks)** | **D** | **AO2** | **Funct.**

5 An architect has ordered roof trusses to be made for a roof.
This is a sketch of the outline of the roof.

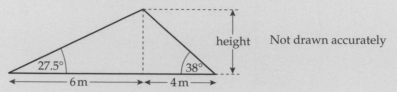

Not drawn accurately

Use a scale drawing to show that the height of the roof is about 3.1 m.   **(3 marks)** | **D** | **AO2** | **Funct.**

6 A solution to the equation $2x^3 - 8x = 162$ lies between 4 and 5.
Use the method of trial and improvement to find this solution correct to 1 decimal place.   **(4 marks)** | **C** | **AO1**

7 Calculate the perimeter of a semicircle with diameter 8 cm.
Leave your answer in terms of π.   **(2 marks)** | **C** | **AO2**

8 Make a copy of the line RT, exactly 8 cm long.

R———————————————————T

Find and shade the region of points that satisfy both of the following conditions.
① The points are nearer to R than T
② The points are not further than 5 cm from T.   **(3 marks)** | **C** | **AO2**

9 a On a coordinate grid, draw the graph of $y = x^2 + 2x - 1$ for $-4 \leqslant x \leqslant 2$.
Draw the x-axis going from -4 to $+2$ and the y-axis going from -4 to $+10$.   **(3 marks)** | **C**
   b Use your graph to solve the equation $x^2 + 2x - 1 = 1$ **(2 marks)** | **C**

10* This triangular prism has a total surface area of 152 cm².

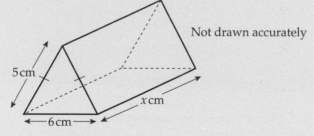

Not drawn accurately

What is the length, x cm, of the prism?
You **must** show your working.   **(3 marks)** | **B** | **(2 marks)** | **C** | **AO3**

11 a Triangle ABC is right angled at B.
$AC = 12$ cm and $\angle ACB = 42°$

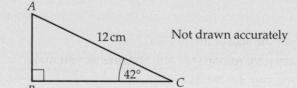

Not drawn accurately

Show, by calculation that the length of AB is 8.0 cm correct to 1 decimal place.   **(2 marks)** | **B** | **AO2**

   b Triangle PQR is right angled at Q.
$PQ = 25$ m and $QR = 35$ m.
Calculate the size of $\angle QPR$.

Not drawn accurately

Give your answer correct to 2 significant figures.   **(3 marks)** | **B** | **AO1**

12 A building is 40 m tall. It casts a shadow of length 48 m at midday.
A tree next to the building casts a shadow of length 18 m at midday.
Calculate the height of the tree.   **(3 marks)** | **B** | **AO2**

13 Mr Smith buys 2 adult and 3 child tickets for a boat trip. He pays a total of £101.50.
Mrs Singh pays £161 for 3 adult tickets and 5 child tickets.
What is the cost of an adult ticket and the cost of a child ticket?   **(4 marks)** | **B** | **Funct.**

14 The diagram shows a cone.

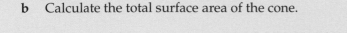

a Calculate the volume of the cone.   **(2 marks)** | **A** | **AO1**
b Calculate the total surface area of the cone.   **(3 marks)** | **A** | **AO1**

15 Triangle *RST* has *RS* = 25 cm, *RT* = 16 cm and ∠*RST* = 35°.

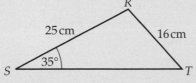

Not drawn accurately

Calculate the length of *ST*.   **(4 marks)** | **A*** | **AO2**

16* a *O* is the centre of the circle.
EF is a tangent to the circle at *D*.
Angle *CAD* = 32°.

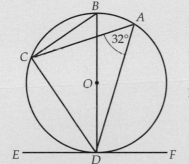

Not drawn accurately

Work out the size of angle *ODC*.
You must give reasons for any statements you make.   **(2 marks)** | **A** | **AO1**

b *O* is the centre of the circle.
PQRS is a cyclic quadrilateral.
Angle *POR* = 2*x*°.

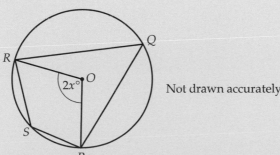

Not drawn accurately

Prove that angle *RQP* + angle *RSP* = 180°.   **(3 marks)** | **A*** | **AO3**

17 This is the graph of $y = x^2 - 2x - 3$.

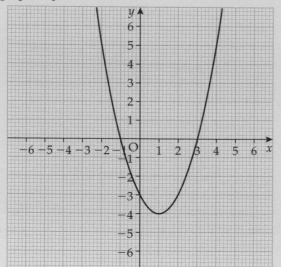

x	-2	-1	0	1	2	3	4
y	5	0	-3	-4	-3	0	5

Use the table of coordinates to make a copy of the graph above.

By drawing an appropriate linear graph, solve the equation $x^2 - 3x - 5 = 0$ **(4 marks)** | A*

18 The diagram shows a cuboid with a square cross section.

The length of the cuboid is twice the height.

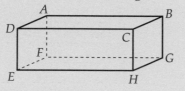

Not drawn accurately

Calculate the angle between *BE* and the base *EFGH*. **(5 marks)** | A* | AO3

1* **a** There are 50 members in a scuba diving club.

22 of the members are women.

What percentage of the members are men?   **(3 marks)** | **D**

b The rate of VAT was raised in January 2010 from 15% to $17\frac{1}{2}$%.

Special offer!
Cement mixer £380 + VAT

Work out the difference in price of a cement mixer due to the increase in VAT.

(3 marks) | **D** | **AO2** | **Funct.**

2 Enid is travelling to Cumbria for her holiday.

She begins her journey at 9 am and travels 80 km in the first hour.

Between 10 am and 11 am she travels only 50 km due to heavy traffic.

At 11 am she stops for a half-hour break.

She reaches her destination after a further $2\frac{1}{2}$ hours and a distance of 150 km.

a Draw a distance–time graph of Enid's journey.   **(4 marks)** | **D** | **AO2**

b During which section of her journey was Enid travelling the fastest?

Explain how you know.   **(1 mark)** | **D**

c Work out Enid's average speed for the whole journey.   **(3 marks)** | **D** | **Funct.**

3 A shoe shop records the number of each size shoe it sells.

These are the results for one week.

Shoe size	Frequency
4	1
5	14
6	17
7	0
8	8

a Write down the range of this data.   **(1 mark)** | **D**

b Write down the mode of this data.   **(1 mark)** | **D**

c Calculate the mean shoe size sold.   **(3 marks)** | **D**

4 Here is a trapezium.

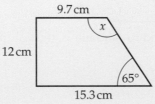

a State the value of x.   **(1 mark)** | **D**

b Work out the area of the trapezium.   **(2 marks)** | **D**

5* John is a builder. He mixes his own concrete out of sand and cement.

The table shows the sand : cement ratios for different types of concrete.

Type of concrete	Sand : cement
general building (above ground)	5 : 1
general building (below ground)	3 : 1
internal walls	8 : 1

John is starting a new job. He estimates that he needs 240 kg of concrete for general building above ground and 180 kg of concrete for internal walls.

Sand and cement are both sold in 25 kg bags.

Work out how many bags of sand and how many bags of cement John needs to buy.

(6 marks) | **C** | **AO2** | **Funct.**

6 The table shows the distances, d km, that some people cycle to work.

Distance, d (km)	Frequency
$0 \leqslant d < 5$	13
$5 \leqslant d < 10$	9
$10 \leqslant d < 15$	5
$15 \leqslant d < 20$	2
$20 \leqslant d < 25$	3

    a Which class interval contains the median?

        Explain how you worked out your answer. **(2 marks)** | C

    b Explain why it is not possible to calculate the exact mean distance. **(1 mark)** | C | AO2

7 Dylan starts with this shape.   He transforms the shape to make this pattern.

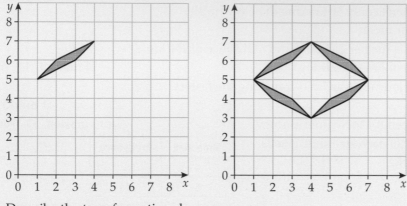

    Describe the transformations he uses. **(4 marks)** | C | AO3

8* The diagram shows a solid. The lengths x, y and z are shown.

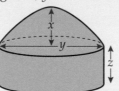

    One of the following formulae may be used to estimate V, the volume of the solid.

$$V = 4x + 2y + 3z$$
$$V = 4x^2y + 2x^2z$$
$$V = 2x(3y^2 + 4z)$$

    a Explain why the formula $V = 4x + 2y + 3z$ cannot be used to estimate the
        volume of the solid. **(1 mark)** | B

    b State, with a reason, which of the above formulae may be used to estimate
        the volume of the solid. **(2 marks)** | B

9* a Factorise $x^2 - 49$ **(2 marks)** | B

    b **i** Show that $(x + y)^2 - (x - y)^2 \equiv 4xy$ **(2 marks)** | B | AO2

       ii Use the identity in part **i** to work out $23^2 - 17^2$ **(2 marks)** | B | AO2

    c Prove that the sum of three consecutive integers is a multiple of 3. **(3 marks)** | B | AO3

10 A report into use of plastic bags in the UK states that approximately 7 800 million
    plastic bags are used in the UK each year.

    a Write this number in standard form. **(1 mark)** | B

    b The population of the UK is approximately 6×10^7.

        What is the mean number of plastic bags used per person in the UK? **(2 marks)** | B | Funct.

11 *AC* is a tangent to the circle at *B*.
Angle *EOD* = 210°
Angle *EDB* = 48°.

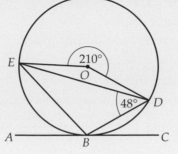

   a   Give a reason why angle *EBD* = 105°.   **(1 mark)** | A
   b   Work out the value of angle *DBC*.   **(2 marks)** | A

12 The diagram shows a major segment of a circle of radius 15 cm.
The length of the major arc is 27π cm.

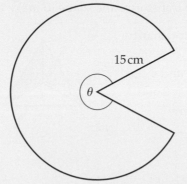

   Calculate the angle of the major segment, marked θ on the diagram.   **(3 marks)** | A

13 A is the point (3, −2) and B is the point (6, 10).
Work out the equation of the line parallel to AB that passes through the point (1, 9).   **(4 marks)** | A | AO3

14 Simplify $64^{-\frac{2}{3}}$ **(3 marks)** | A*

15 **a**   Write the expression $x^2 - 6x - 23$ in the form $(x + p)^2 + q$.   **(2 marks)** | A*
   b   Hence solve the equation $x^2 - 6x - 23 = 0$.
           Give your answer in the form $a \pm b\sqrt{2}$.   **(3 marks)** | A*

16 A bag contains three white and four red discs.
Anil takes a disc at random from the bag. He does not replace it.
Bern then takes a disc at random from the bag.
Anil wins if both the discs are the same colour.
Who has a better chance of winning?
You must show working to support your answer.   **(3 marks)** | A* | AO3

Linear Paper 2 Higher tier   2 hr, Calculator allowed

1 Andy and Bari share £240 in the ratio 1 : 3.
How much does Bari get?   **(3 marks)** | D

2 Steven goes shopping. He has £80 to spend.
He spends £24.50 on a pair of squash shoes and £6.99 on some squash balls.
He sees this advert for a squash racket.

> **Squash Racket:** normal price £95
> **Special offer!**
> 45% off normal price

Does Steven have enough money to buy the squash racket?
You **must** show your working.   **(4 marks)** | D | AO2 | Funct.

3 a Use your calculator to work out $\sqrt[3]{117}$.

  i Write down your full calculator display. **(1 mark)** | **D**

  ii Write down your answer correct to one decimal place. **(1 mark)** | **D**

b Use your calculator to work out $\dfrac{9.49}{1.88 + 2.27}$

  i Write down your full calculator display. **(1 mark)** | **D**

  ii Write down your answer to a suitable degree of accuracy. **(1 mark)** | **D**

4 This is the area of a garden that is going to have turf laid to make a new lawn.

Turf costs £2.25 per square metre.

It can only be bought in whole numbers of square metres.

Work out the cost of the turf for the new lawn.

You **must** show your working.

(4 marks) | **D** | **AO2** | **Funct.**

5* Megan designs a game to raise money for charity.

She makes two fair spinners.

One is four-sided and numbered 1, 2, 3 and 4.

The other is three-sided and numbered 1, 2 and 3.

Players spin the spinners at the same time and multiply the numbers together.

Each player pays £1 to play the game.

If a player gets a score of 8 they win £3.

If a player gets a score less than 3 they win £2.

If 300 people play the game, how much money should Megan expect to make?

Show clearly how you worked out your answer. **(6 marks)** | **D** | **AO3**

6 a Factorise $6y + 18$ **(1 mark)** | **D**

b Solve $\dfrac{x}{7} = -8$ **(1 mark)** | **D**

c $d \triangle e$ means $9d + 7e$

  When $x \triangle 4 = 6 \triangle x$, work out the value of x. **(3 marks)** | **D** | **AO2**

d Expand and simplify $(x - 7)(x + 3)$ **(2 marks)** | **C**

7 Maria runs a business making tin containers.

She uses this formula to work out the approximate total surface area of a cylindrical tin.

> $A = 6r(r + h)$ where: A is the total surface area in cm^2
> r is the radius of the tin in cm
> h is the height of the tin in cm

a Use the formula to work out the approximate total surface area of a cylindrical tin with radius 3 cm and height 9 cm. **(2 marks)** | **D** | **Funct.**

b Rearrange the formula to make h the subject. **(2 marks)** | **B** | **Funct.**

8 *ABCD* is a quadrilateral.

Angle *D* is 90°

Angle $A = x$, angle $B = 3x - 20°$ and angle $C = 2x + 32°$

Work out the largest angle in the quadrilateral.

You **must** show your working.

Not drawn accurately

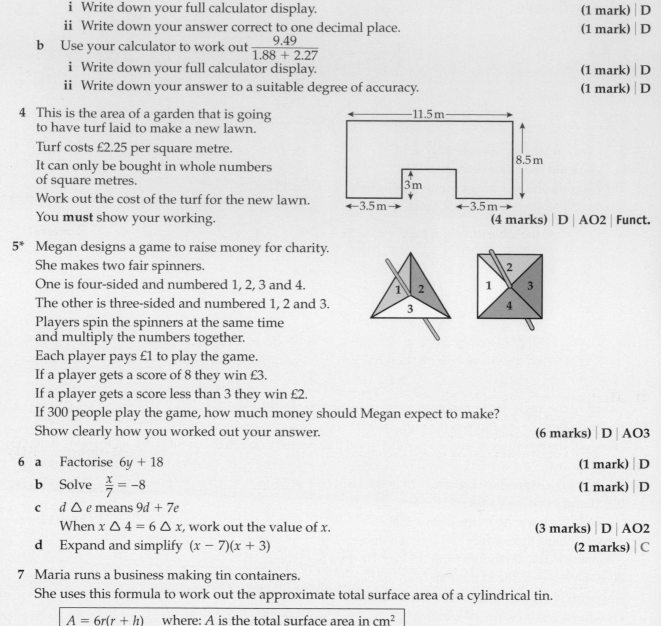

(5 marks) | **D** | **AO3**

9 Harry is going to carry out a survey on visits to the dentist.

These are two of the questions he has written.

> 1. Do you agree that it takes to long to get am appointment to see the dentist?
>
>   Strongly agree ☐   Agree ☐   Don't know ☐
>
> 2. How often do you visit the dentist?

   a Give two reasons why question 1 is unsuitable. **(2 marks)** | C

   b Give a suitable response section for question 2. **(1 mark)** | C

   c Harry decides to carry out his survey outside his local dentist surgery.
     Explain why this sample is likely to be unrepresentative. **(1 mark)** | C | AO2

10 a Complete the table of values for $y = x^2 - 2x - 4$. **(2 marks)** | C

x	−3	−2	−1	0	1	2	3	4
y		4	−1	−4		−4	−1	

   b Draw a coordinate grid that goes from −3 to +4 on the x-axis and −5 to +15 on the y-axis.
     On the grid, draw the graph of $y = x^2 - 2x - 4$ for values of x from −3 to +4. **(2 marks)** | C

   c Use your graph to write down the solutions to the equation $x^2 - 2x - 4 = 0$ **(2 marks)** | C

11 M is the point (5, 6) and N is the point (9, 14).

   a Work out the midpoint of the line MN. **(2 marks)** | C

   b Calculate the length of the line MN.
     Give your answer correct to one decimal place. **(2 marks)** | C | AO2

12* a Show that $4(4y - 7) - 5(2y - 9) \equiv 3(2y + 7) - 4$ **(4 marks)** | C | AO3

    b Solve the equation $\dfrac{3x + 4}{2} - \dfrac{2x + 1}{3} = 5$ **(4 marks)** | B

13* A company produce two types of light bulb, type A and type B.

The lifetime, in hours, of a sample of 80 of each type of bulb was measured.

The cumulative frequency diagram shows the results for the type A light bulb.

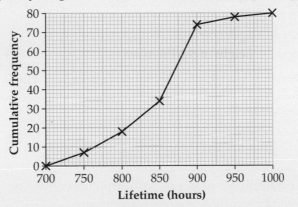

   a Estimate the median and interquartile range for the type A light bulb. **(3 marks)** | B | Funct.

   b The median and interquartile range for the type B light bulb are 825 hours and 110 hours respectively.
     Which type of bulb is longer lasting?
     Use the medians and inter-quartile ranges to clearly explain your answer. **(2 marks)** | B | AO2 | Funct.

14 A catering company cooks meals for parties.

They offer three main courses: lasagne (L), fish (F) or quiche (Q).

To accompany the main course they offer either salad (S) or chips (C).

The company use previous data to estimate the number of different types of meals they need to cook. The probability of a person choosing lasagne is 0.4 and fish is 0.5.

The probability of a person choosing salad is 0.25.

a Copy and complete the tree diagram to show all the possible outcomes.

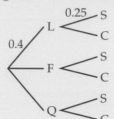

(2 marks) | B | AO2 | Funct.

b Work out the probability that a person chooses fish and chips.   (2 marks) | B | Funct.

c At the next party, 200 guests are expected.

Estimate the number of quiche and salad meals the company will need to cook.

(2 marks) | B | AO2 | Funct.

15 The diagram shows some angles around a point.

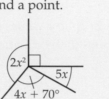

a Show that x satisfies the equation $2x^2 + 9x - 200 = 0$ (2 marks) | B | AO2

b Solve the equation $2x^2 + 9x - 200 = 0$.

Hence work out the sizes of the angles around the point.   (4 marks) | A

16 Serge wants to replace the old tiles on his garage roof with new clay tiles.

The recommended minimum angle of elevation of a roof suitable for clay tiles is 35°.

The diagram shows the dimensions of his roof.

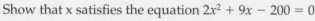

Should Serge use clay tiles on his garage roof?

You **must** show working to support your answer.   (3 marks) | A | AO2 | Funct.

17 A football club has 28 000 season ticket holders.

They are classified by age as follows.

Age (years)	Under 21	21–40	41–60	Over 60
Number of season ticket holders	6440	12 680	7250	1630

The football club wants to take a stratified sample of 500 season ticket holders.

Calculate the number that should be sampled from each age group.   (3 marks) | A | Funct.

18 Two cones are mathematically similar.

The smaller cone has a volume of 120 cm³.

The larger cone has a volume of 405 cm³.

The curved surface area of the smaller cone is 102 cm².

Work out the curved surface area of the smaller cone.   (4 marks) | A | AO3

19 Will draws a circle at two of the opposite vertices of a rhombus.

The rhombus has a side length of 6 cm.

The diagram shows the shape Will has drawn.

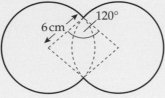

a Show that the perimeter of the shape is 20π cm. **(3 marks)** | A | AO2

b Work out the area of the shape. **(5 marks)** | A* | AO3

20 OAB is a triangle with $\overrightarrow{OA} = \mathbf{a}$ and $\overrightarrow{OB} = \mathbf{b}$.

M is the mid-point of OB and P is the point on AB such that $AP:PB = 1:2$.

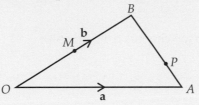

Find expressions for these vectors, in terms of $\mathbf{a}$ and $\mathbf{b}$, simplifying your answers when possible.

a $\overrightarrow{AB}$ **(1 mark)** | A

b $\overrightarrow{OM}$ **(1 mark)** | A

c $\overrightarrow{OP}$ **(2 marks)** | A*

d $\overrightarrow{MP}$ **(2 marks)** | A*

21 Solve these simultaneous equations using an algebraic method.

$$y - 5x = 2$$
$$y = x^2 + x - 10$$

(5 marks) | A*